Facts and Mysteries
in
Elementary Particle Physics

Facts and Mysteries
in
Elementary Particle Physics

u u u up

d d d down

ν_e e-neutrino

e electron

Martinus J. G. Veltman

MacArthur Emeritus Professor of Physics
University of Michigan, Ann Arbor, USA
and
NIKHEF, Amsterdam, the Netherlands

World Scientific
New Jersey • London • Singapore • Hong Kong

Published by

World Scientific Publishing Co. Pte. Ltd.

5 Toh Tuck Link, Singapore 596224

USA office: Suite 202, 1060 Main Street, River Edge, NJ 07661

UK office: 57 Shelton Street, Covent Garden, London WC2H 9HE

Library of Congress Cataloging-in-Publication Data
Veltman, Martinus.
    Facts and mysteries in elementary particle physics / Martinus J.G. Veltman.
      p. cm.
    Includes index.
    ISBN 981-238-148-1 -- ISBN 981-238-149-X (pbk.)
     1. Particles (Nuclear physics) I. Title.

QC793.2.V45 2003
539.7'2--dc21
                                     2003042273

British Library Cataloguing-in-Publication Data
A catalogue record for this book is available from the British Library.

Printed by FuIsland Offset Printing (S) Pte Ltd, Singapore

Table of Contents

Introduction

The twentieth century has seen an enormous progress in physics. The fundamental physics of the first half of that century was dominated by the theory of relativity, Einstein's theory of gravitation, and the theory of quantum mechanics. The second half of the century saw the rise of elementary particle physics. In other branches of physics much progress was made also, but in a sense developments such as the discovery and theory of superconductivity are developments in width, not in depth. They do not affect in any way our understanding of the fundamental laws of Nature. No one working in low-temperature physics or statistical mechanics would presume that developments in those areas, no matter how important, would affect our understanding of quantum mechanics.

Through this development there has been a subtle change in point of view. In Einstein's theory of gravitation space and time play an overwhelming, dominant role. The movement of matter through space is determined by the properties of space. In this theory of gravitation matter defines space, and the movement of matter through space is then determined by the structure of space. A grand and imposing view, but despite the enormous authority of Einstein most physicists no longer adhere to this idea. Einstein spent the latter part of his life trying to incorporate electro-magnetism into this picture, thus trying to describe electric and magnetic fields as properties of space-time. This became known as his quest for a unified theory. In this he really never succeeded, but he was not a man given to abandon easily a point of view.

Max Planck (1858–1947), founder of quantum physics. In 1900 he conceived the idea of quantized energy, introducing what is now called Planck's constant, one that sets the scale for all quantum phenomena. In 1918 he received the Nobel prize in physics. Citation: "In recognition of the services he rendered to the advancement of Physics by his discovery of energy quanta." Planck was one of the first to recognize Einstein's work, in particular the theory of relativity. According to Einstein, Planck treated him as something like a rare stamp. Well, in any case Planck got Einstein to Berlin.

Planck's importance and influence cannot be overstated. It is very just that the German Max Planck Society is named after him. He is the very initiator of quantum mechanics. Discrete structures (atoms) had been suggested before Planck, but he deduced quantum behaviour for an up to then continuous variable, energy. He did it on the basis of a real physical observation.

Planck had other talents beyond physics. He was a gifted pianist, composed music, performed as a singer and also acted on the stage. He wrote an opera "Love in the Woods" with "exciting and lovely songs".

His long life had a tragic side. His first wife died in 1909, after 22 years of marriage, leaving him with two sons and two daughters. The oldest son was killed in action in World War I, and both of his daughters died quite young in childbirth (1918 and 1919). His house was completely destroyed in World War II; his youngest son was implicated in the attempt made on Hitler's life on July 20, 1944 and was executed in a gruesome manner by Hitler's henchmen.

However, his view became subsequently really untenable, because next to gravitation and electromagnetism other forces came to light. It is not realistic to think that these can be explained as properties of space-time. The era of that type of unified theory is gone.

The view that we would like to defend can perhaps best be explained by an analogy. To us space-time and the laws of quantum mechanics are like the decor, the setting of a play. The elementary particles are the actors, and physics is what they do. A door that we see on the stage is not a door until we see an actor going through it. Else it might be fake, just painted on.

Thus in this book elementary particles are the central objects. They are the actors that we look at, and they play a fascinating piece. There are some very mysterious things about this piece. What would you think about a play in which certain actors always occur threefold? These actors come in triples, they look the same, they are dressed completely the same way, they speak the same language, they differ only in their sizes. But then they really do differ: one of the actors is 35,000 times bigger than his otherwise identical companion! That is what we see today when systematizing elementary particles. And no one has any idea why they appear threefold. It is the great mystery of our time. Surely, if you saw a play where this happened you would assume there had to be a reason for this multiplicity. It ought to be something you could understand at the end of Act One. But no. We understand many things about particles and their interactions, but this and other mysteries make it very clear that we are nowhere close to a full understanding. And, most important: we still do not understand gravity and its interplay with quantum mechanics.

This book has been set up as follows. Chapter 1 contains some preliminaries: atoms, nuclei, protons, neutrons and quarks are introduced, as well as photons and antiparticles. Furthermore there is an introductory discussion of mass and energy, followed by a description of the notion of an event, central in particle physics. The Chapter closes with down-to-earth type subjects such as units used and particle naming. We begin in Chapter 2 by introducing the actors, the elementary particles and their interactions. Forces are understood today as due to the interchange of particles, and therefore we will use the word 'interactions' rather than the word 'forces'. The ensemble of particles and forces described in Chapter 2 is known as the Standard Model. In Chapter 3 some very elementary concepts of quantum mechanics shall be discussed, and in Chapter 4 some of the aspects of ordinary mechanics and the theory of relativity. In other words, we must also discuss the stage on which the actors appear. An overview of the basic ideas and experimental methods in Chapters 5 and 6 will make it clear how research in this domain is organized and progresses. Chapter 7 contains an overview of the 1963 CERN neutrino experiment, showing how these things work in reality. It shows how the simple addition of one more entry in the table of known elementary particles is based on colossal experimental efforts. In Chapter 8 the observed particle spectrum (including bound states), called the particle zoo, will be reviewed, showing how the idea of quarks came about. That idea reduced the observed particle zoo to a few basic elementary particles. In Chapter 9 we come to the more esoteric part: the understanding of the theory of elementary particles. Chapter 10 contains a further discussion of the Higgs particle and the experimental search for it. Finally, in Chapter 11 a short description of the theory of strong interactions will be presented. The strong interactions are responsible for the forces between the quarks, giving rise to the particle zoo, the complex spectrum of particles as mentioned above.

There is one truth that the reader should be fully aware of. Trying to explain something is a daunting endeavour. You cannot explain the existence of certain particles much as you cannot explain the existence of this Universe. In addition, the laws of quantum mechanics are sufficiently different from the laws of Newtonian mechanics which we experience in daily life to cause discomfort when studying them. Physicists usually cross this barrier using mathematics: you understand something if you can compute it. It helps indeed if one is at least capable of computing what happens in all situations. But we cannot assume the reader to be familiar with the mathematical methods of quantum mechanics, so he will have to swallow strange facts without the support of equations. We can only try to make it as easy as possible, and shall in any case try to state clearly what must be swallowed!

Acknowledgments

Many people have helped in the making of this book, by their criticism and constructive comments. I may single out my daughter Hélène, who has gone more than once through the whole book. Special mention needs to be made of Karel Mechelse, himself a neurologist, who read through every Chapter and would not let it pass if he did not understand it. I am truly most grateful to him. If this book makes sense to people other than particle physicists then that is his merit. Furthermore I would like to mention the help of Val Telegdi, untiring critic of both physics and language with a near perfect memory. I really profited immensely from his comments. I cannot end here without mentioning the wonderful two-star level dinners that his wife Lia prepared at their home; they compensated in a great way for the stress of undergoing Val's criticism.

Thanks are also due to several people at the NIKHEF (Nationaal Instituut voor Kernfysica en Hoge Energie Fysica, the Dutch particle physics institute), especially Kees Huyser who knows everything about computers, pictures and typesetting.

Further Reading

There are many books about physics, on the popular and not so popular level and each has its particular virtues. Two books deserve special mention:

A. Pais: *Subtle is the Lord... The Science and the Life of Albert Einstein*, Oxford University Press 1982, ISBN 0-19-853907-X.

A. Pais: *Inward Bound. Of Matter and Forces in the Physical World*, Oxford University Press 1986, ISBN 0-19-851997-4.

These two books, masterpieces, contain a wealth of historical data and an authoritative discussion of the physics involved. We have extensively consulted them and occasionally explicitly quoted from them. One remark though: Pais was a theoretical physicist and his books are somewhat understating the importance of experiments as well as of experimental ingenuity. Progress almost always depends on experimental results, without which the smartest individual will not get anywhere. For example, the theory of relativity owes very much to the experiments of Michelson concerning the speed of light. And Planck came to his discovery due to very precise measurements on blackbody radiation done in the same place, Berlin, in which he was working. On the other hand, to devise useful experiments an experimental physicist needs some understanding of the existing theory. It is the combination of experiment and theory that has led to today's understanding of Nature.

A book written by an experimental physicist:

L. Lederman: *The God Particle*, Houghton Mifflin Company, Boston, New York 1993, ISBN 0-395-55849-2.

Thumbnail Sketches

There are in this book many short sketches, or vignettes as I call them, with pictures. I would like to state clearly that these vignettes must not be seen as a way of attributing credit to the physicists involved. Many great physicists are not present in the collection. The main purpose is to give a human face to particle physics, not to assign credit. The fact that some pictures were easier to obtain than others has played a role as well.

Equations

Sometimes slightly more mathematically oriented explanations have been given. As a rule they are not essential to the reasoning, but it may help. Such non-essential pieces are set in smaller type on a shaded background.

1

Preliminaries

1.1 Atoms, Nuclei and Particles

All matter is made up from molecules, and molecules are bound states of atoms. For example, water consists of water molecules which are bound states of one oxygen atom and two hydrogen atoms. This state of affairs is reflected in the chemical formula H_2O.

There are 92 different atoms seen in nature (element 43, technetium, is not occurring in nature, but it has been man-made). Atoms have a nucleus, and electrons are orbiting around these nuclei. The size of the atoms (the size of the outer orbit of the electrons) is of the order of $1/100\,000\,000$ cm, the nucleus is $100\,000$ times smaller. The atom is therefore largely empty. Compare this: suppose the nucleus is something like a tennis ball (about 2.5 inch or 6.35 cm diameter). Then the first electron circles at a distance of about 6.35 km (4 miles). It was Rutherford, in 1911, who discovered that the atom was largely empty by shooting heavy particles (α particles,[a] emanating from certain radioactive materials) at nuclei. These relatively heavy particles ignored the very light circling electrons much like a billiard ball would not notice a speck of dust. So they scattered only on the nucleus. Without going into detail we may mention that Rutherford actually succeeded in estimating the size of the nucleus.

---

[a]An α particle is nothing else but a helium nucleus, that is a bound state of two protons and two neutrons. That was of course not known at the time.

Niels Bohr (1885–1962). In 1913 he proposed the model of the atom, containing a nucleus orbited by electrons. In the period thereafter he was the key figure guiding the theoretical development of quantum mechanics. While Heisenberg, Schrödinger, Dirac and Born invented the actual mathematics, he took it upon himself to develop the physical interpretation of these new and spooky theories. Einstein never really accepted it and first raised objections at the Solvay conference of 1927. This led to the famous Bohr-Einstein discussions, where the final word (at the 1930 Solvay conference) was Bohr's, answering Einstein using arguments from Einstein's own theory of gravitation. Even if Bohr had the last word, Einstein never wavered from his point of view.

It should be mentioned that Bohr started his work leading to his model at Manchester, where Rutherford provided much inspiration. Bohr's famous trilogy of 1913, explaining many facts, in particular certain spectral lines of hydrogen (Balmer series), may be considered (in Pais' words) the first triumph of quantum dynamics.

Bohr received the Nobel prize in 1922. In World War II, after escaping from Denmark, he became involved in the American atomic bomb project. After the war he returned to Copenhagen, and as a towering figure in Europe he played an important role in the establishment of CERN, the European center for particle physics. In fact he became the first director of the theory division, in the beginning temporarily located at his institute in Copenhagen.

The nucleus contains protons and neutrons, also called nucleons. The proton has an electric charge of +1 (in units where the charge of the electron is −1), the neutron is electrically neutral. The number of electrons in an atom equals the number of protons in the nucleus, and consequently atoms are electrically neutral. It is possible to knock one or more electrons off an atom; the remainder is no longer electrically neutral, but has a positive charge as there is then an excess number of protons. Such an object is called an ion, and the process of knocking off one or more electrons is called ionization. For example electric discharges through the air do that, they ionize the air.

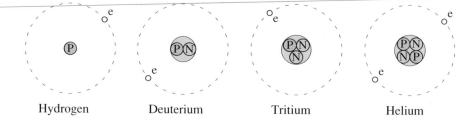

| Hydrogen | Deuterium | Tritium | Helium |

The lowest mass atom is the hydrogen atom, with one electron and a nucleus consisting of just one proton. The nucleus of heavy hydrogen, called deuterium, has an extra neutron. If both hydrogen atoms in a water molecule are deuterium atoms one speaks of "heavy water". In natural water one finds that about 0.015% of the molecules contains one or two deuterium atoms. Tritium is hydrogen with two extra neutrons in the nucleus. Helium is the next element: two electrons and a nucleus containing four nucleons, i.e. two protons and two neutrons.

Nuclear physics is that branch of science that covers the study of atomic nuclei. The nuclear experimenter shoots electrons or other projectiles into various nuclei in order to find out what the precise structure of these nuclei is. He is not particularly interested in the structure of the proton or neutron, although nowadays the boundary between nuclear physics and elementary particle physics is becoming blurred.

Ernest Rutherford (1871–1937). He investigated and classified radioactivity. He did the first experiments exhibiting the existence of a nucleus. In 1908 he received the Nobel prize in chemistry, "for his investigations into the disintegration of the elements, and the chemistry of radioactive substances". He is surely one of the rarest breed of people, doing his most important work after he received the Nobel prize. I am referring here to the scattering of alpha particles from nuclei. The actual experiment was done by Geiger (of the Geiger-Müller counter, actually initiated by Geiger and Rutherford) and Marsden, under the constant influence of Rutherford. Later, Rutherford produced the relevant theory, which is why we speak today of Rutherford scattering.

He was the first to understand that there is something peculiar about radioactivity. Anyone listening to a Geiger counter ticking near a radioactive source realizes that there is something random about those ticks. It is not like a clock. That was the first hint of the undeterministic behaviour of particles. Rutherford noted that.

Rutherford was a native of New Zealand. He was knighted in 1914 and later became Lord Rutherford of Nelson. His importance goes beyond his own experimental work. His laboratory, the Cavendish (built by Maxwell), was a hotbed of excellent physicists. Chadwick discovered the neutron there (Nobel prize 1935) and in 1932 Cockcroft and Walton (Nobel prize 1951) constructed a 700 000 Volt generator to make the first proton accelerator. Some laboratory!

A proton, as we know now, contains three quarks. There are quite a number of different quarks, with names that somehow have come up through the years. There are "up quarks" (u) and "down quarks" (d), and each of them comes in three varieties, color coded red, green and blue (these are of course not real colors but just a way to differentiate between the quarks). Thus there is a red up quark, a green up quark and a blue up quark, and similarly for the down quark. A proton contains two up quarks and a down quark, all of different colors, while a neutron contains one up quark and two down quarks likewise of different colors. The figures show a symbolic representation of the up and down quarks, and the quark contents of the proton and the neutron. Just to avoid some confusion later on: sometimes we will indicate the color of a quark by a subscript, for example u_r means a red up quark.

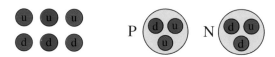

It should be emphasized that while we shall draw the quarks (as well as electrons and others) as little balls, it is by no means implied that they are actually something like that. For all we know they are point-like. No structure of a quark or electron has ever been observed. We just draw them this way so that we can insert some symbol and also color them.

Protons and neutrons can be observed as free particles. For example, if we strip the electron from a hydrogen atom we are left with a single proton. Single neutrons decay after a while (10 minutes on the average), but live long enough to be studied in detail. However, the quarks never occur singly. They are confined, bound within proton or neutron. The way these quarks are bound in a proton or neutron is quite complicated, and not fully understood. Statements about the quark content of proton and neutron must be taken with a grain of salt, because in addition there are particles called gluons which cause the binding and which are much more dominantly present than for example photons in an atom (the atomic binding is due to electromagnetic forces, thus photons do the job of binding the electrons to the nucleus). In fact, much of the mass of a proton or a neutron resides in the form of energy of the gluons, while the energy residing in the electric field of an atom is very small.

For all we know electrons and quarks are elementary particles, which means that in no experiment has there anything like a structure of these particles been seen. They appear point-like, unlike the proton, neutron, nucleus and atom that have sizes that can be measured. It is of course entirely possible that particles that are called elementary today shall turn out to be composite; let it be said though that they have been probed quite extensively. This book is about elementary particles. The aim is to know all about them, their properties and their interactions. The idea is that from this nuclear physics, atomic physics, chemistry, in fact the whole physical world derives. Thus particles and their interactions are the very fundamentals of nature. That is the view now. An elementary particle physicist studies primarily these elementary particles and not the larger structures such as protons, nuclei or atoms.

The main laboratory for elementary particle research in Europe has been named CERN (Conseil Européen pour la Recherche Nucléaire), now officially called European Organization for Nuclear Research and that is a misnomer. In principle no nuclear physics is being done there. In the days (1953) when CERN came into being nuclear physics was a magic word if money was needed! Strangely enough, the organization called Euratom is one that studies nuclei and not atoms. Another important laboratory is DESY, Deutsches Elektronen-Synchrotron, in Hamburg, Germany. In the US there are several laboratories, among them BNL, Brookhaven National Laboratory (at Long Island near New York), Fermi National Laboratory (near Chicago) and SLAC, Stanford Linear Accelerator Center (near San Francisco).

Papers that changed the world: Planck's quantum.
Verh. Deutsch. Phys. Ges. 2 (1900) 237

Zur Theorie des Gesetzes
der Energieverteilung im Normalspectrum;
von M. Planck.

(Vorgetragen in der Sitzung vom 14. December 1900.)

(Vgl. oben S. 235.)

M. H.! Als ich vor mehreren Wochen die Ehre hatte, Ihre Aufmerksamkeit auf eine neue Formel zu lenken, welche

~~~

teilung auf unendlich viele Arten möglich. Wir betrachten aber — und dies ist der wesentlichste Punkt der ganzen Berechnung — E als zusammengesetzt aus einer ganz bestimmten Anzahl endlicher gleicher Teile und bedienen uns dazu der Naturconstanten $h = 6{,}55 . 10^{-27}$ [erg $\times$ sec]. Diese Constante mit der gemeinsamen Schwingungszahl ν der Resonatoren multiplicirt ergiebt das Energieelement ε in erg, und durch

In this paper Planck tries to find an explanation of his successful formula for blackbody radiation. He succeeds in that by introducing energy quanta and he proposes (in words) the equation $\epsilon = h\nu$. The modern value for h is 6.626×10^{-27}. Surprisingly close!

On the theory of the Energy Distribution Law of the Normal Spectrum
by M. Planck

Gentlemen: when some weeks ago I had the honour to draw your attention to a new formula...

~~~

... We consider however — this is the most essential point of the whole calculation — E to be composed of a well-defined number of equal parts and use thereto the constant of nature $h = 6.55 \times 10^{-27}$ erg sec. This constant multiplied by the common frequency ν of the resonators gives us an energy element ϵ in erg, and ...

1.2 Photons

In 1905 Einstein proposed the daring idea that electromagnetic radiation is quantized and appears only in precisely defined energy packets called photons. It took 15 years before this idea was accepted and initially it was considered by many as a bad mistake. But in 1921 Einstein was awarded the Nobel prize in physics and the quotation of the Swedish Academy stated that this prize was awarded because of his services to Theoretical Physics, and in particular for this discovery. Especially the part of Einstein's paper on the photoelectric effect contained barely any mathematics, but it was nevertheless really a wonderful piece of physics. Great physics does not automatically imply complicated mathematics!

When we think of a ray of light we now think of a stream of photons. The energy of these photons depends on the type of electromagnetic radiation; the photons of radio waves have lower energy than those of visible light (in which red light photons are less energetic than blue light photons), those of X-rays are of still higher energy, and gamma rays consist of photons that are even more energetic than those of X-rays. In particle physics experiments the photon energies are usually very high, and one deals often with individual photons. The energy of those photons is more than $100\,000\,000\,000$ times that of the photons emitted by mobile phones. The energy of a photon for a given type of radiation can be computed using a relation published earlier (in 1900) by Planck and involving a new constant that is now called Planck's constant. Planck was the first to introduce quantization, but he did not go so far as to say that light is quantized. He thought of emission in packets, but not that light could exist only in such packets. His hypothesis was on the nature of the process of emission, not on the nature of the radiated light. It seems a small step, but it is precisely this type of step that is so difficult to make.

It is interesting to quote here the recommendation made by Planck and others when nominating Einstein for the Prussian

James Clerk Maxwell (1831–1879). This man wrote down the laws of electromagnetism, and explained light as electromagnetic waves. His equations stand till today. His theory has had enormous consequences. From it developed the theory of relativity, and on the practical side the discovery and application of radio waves by Hertz and Marconi. Maxwell must be ranked among the giants of physics such as Newton and Einstein.

The genius of Maxwell did not limit itself to electromagnetism. He also made large contributions to the study of systems containing many particles, such as a gas in a box, containing many, many molecules. He developed an equation describing the velocity distribution of these molecules. That equation is called the Maxwell velocity distribution.

Maxwell also came up with the idea of a demon, capable of selecting molecules. The demon would sit in some vessel near a hole, and allow passage only to fast-moving molecules. Since the temperature of a gas is directly related to the average velocity of the molecules, it follows that the stream coming out of the hole was hotter than the gas inside. Unfortunately there are no such demons!

In 1874 Maxwell became the first director of the Cavendish laboratory at Cambridge. In those days the difference between theorists and experimentalists was not as sharp as today. His successors were J. J. Thomson (from 1879 till 1919) and Rutherford.

Academy in 1913: "In summary, one can say that there is hardly one among the great problems in which modern physics is so rich to which Einstein has not made a remarkable contribution. That he may sometimes have missed the target in his speculations, as, for example, in his hypothesis of light quanta, cannot really be held against him, for it is not possible to introduce really new ideas even in the most exact sciences without sometimes taking a risk."

The photon concept has an important consequence. In 1873 Maxwell introduced equations describing all electric and magnetic phenomena, now called the Maxwell equations. He then suggested that light is a form of electromagnetic fields, a brilliant idea that worked out wonderfully. We conclude that electromagnetic fields are made up from photons. Therefore, electric and magnetic forces must now be assumed to be due to the action of photons. While it is relatively easy to imagine light to be a stream of photons, it is hard to see how an electric field is due to photons. Yet that is the case, although those photons have properties different from those of the photons of light. It is too early to discuss that here, since one requires for that the concept of the mass-shell, discussed in Chapter 4 (the photons of light are "on the mass shell", those of electric and magnetic fields are not).

As a final comment, recall that light behaves as a propagating wave. Interference experiments show this most clearly. Thus somehow particles, in this case photons, can behave as waves. There one may see the origin of quantum mechanics, or wave mechanics as it was called in the old days. Einstein, well aware of all this, somehow never discovered quantum mechanics. This is one of the more astonishing things: why did he not discover quantum theory? Knowing all about the wave theory of light and having introduced the concept of the photon, he never fused these concepts into one theory. According to Pais, Einstein pondered about this problem in a most intensive way in the period 1905–1910. It seems so straightforward now, yet he missed it.

Papers that changed the world: Einstein's photon.
Annalen der Physik 17 (1905) 132

6. Über einen die Erzeugung und Verwandlung des Lichtes betreffenden heuristischen Gesichtspunkt; von A. Einstein.

Zwischen den theoretischen Vorstellungen, welche sich die

Es scheint mir nun in der Tat, daß die Beobachtungen über die „schwarze Strahlung", Photolumineszenz, die Erzeugung von Kathodenstrahlen durch ultraviolettes Licht und andere die Erzeugung bez. Verwandlung des Lichtes betreffende Erscheinungsgruppen besser verständlich erscheinen unter der Annahme, daß die Energie des Lichtes diskontinuierlich im Raume verteilt sei. Nach der hier ins Auge zu fassenden Annahme ist bei Ausbreitung eines von einem Punkte ausgehenden Lichtstrahles die Energie nicht kontinuierlich auf größer und größer werdende Räume verteilt, sondern es besteht dieselbe aus einer endlichen Zahl von in Raumpunkten lokalisierten Energiequanten, welche sich bewegen, ohne sich zu teilen und nur als Ganze absorbiert und erzeugt werden können.

Bern, den 17. März 1905.

On a heuristic point of view concerning the generation and conversion of light
by A. Einstein

Between theoretical ideas that physicists have …

It indeed seems to me that the observations about "blackbody radiation", photoluminescence, the production of cathode rays by ultraviolet light and others concerning the generation and conversion of light can be understood better under the assumption that the energy of light is distributed discontinuous in space. According to the assumption suggested here, the extension in space of light from a point source is not continuously distributed over a larger and larger domain, but it consists of a finite number of localized energy quantums, that move without division and that can only as a whole be absorbed or emitted.

1.3 Antiparticles

It is useful to mention antiparticles here. They shall be discussed more extensively later, but here we wish to state explicitly that while antiparticles may have some properties different from those of the corresponding particles, they are still just "particles". For example, a particle and its antiparticle have exactly the same mass and both fall downwards in the earth's gravitational field. The antiparticle of the electron is called a positron, and it has the same mass as the electron, but the opposite electric charge. That's all. Do not see anything particularly mysterious in antimatter. It is just a name given, one could equally well have spoken of mirror particles. Also, it is a matter of convention which is called the particle and which the antiparticle. One could equally well have called the positron the particle and the electron the antiparticle. That particles and antiparticles may react with each other quite violently is true, but there are many other (violent) reactions that do not particularly differ in principle from electron-positron reactions. For example, at very high energies two protons colliding with each other produces something quite similar to proton–antiproton collisions.

The importance of the concept of antiparticles follows from a law of nature: to each particle there corresponds an antiparticle that has precisely the same mass, and whose other properties are exactly defined with respect to those of the particle. For example, the electric charge has the opposite sign. The law mentioned allows for the possibility that the antiparticle corresponding to a particle be the particle itself. In that special case the charge of the particle must necessarily be zero. The photon is such a particle. It is its own antiparticle.

There is a standard way to denote an antiparticle: by means of a bar above the particle name or symbol. Thus one could write $\overline{\text{electron}}$ and that would mean a positron. And also, to make the point once again, $\overline{\text{positron}}$ means an electron. This convention will be used throughout this book.

Paul Dirac (1902–1984). He succeeded in combining quantum mechanics and the theory of relativity in 1928, and introduced in 1929 the idea of an antiparticle (although not the name, introduced by de Broglie in 1934). Unfortunately only the expert can appreciate Dirac's awesome work. Ehrenfest from Leiden University, himself no mean physicist, termed it "inhuman". To this day, that work has not lost any of its splendor.

Dirac was not a talkative person and of a quite literal mind, to which numerous anecdotes testify. Here is an example. In a seminar someone asked a question: "Professor Dirac, I do not understand that equation." Dirac did not answer, so the chairman intervened and asked Dirac if he would answer the question. To which Dirac replied: "that was not a question, it was a statement."

Dirac is generally considered the founder of field theory. Field theory is the logical development of quantum mechanics, applicable also to processes in which particles are created or annihilated. For example, ordinary quantum mechanics is enough for determining all possible states of an electron in a hydrogen atom; to actually compute the emission of light if an electron drops from some orbit to a lower orbit requires the machinery of field theory. Field theory has come a long way since Dirac; it has developed gradually over the years and culminated (so far) in the gauge theories of elementary particle interactions.

Dirac received the Nobel prize in 1933.

Bound states also have their associated state. For example, a proton contains three quarks, two u and one d quark, and an antiproton simply contains the corresponding antiparticles: two antiup quarks $\bar{u}$ and one antidown quark $\bar{d}$. At CERN antihydrogen has been created: one positron circling an antiproton.

1.4 Mass and Energy

Energy is a very fundamental concept that plays a central role in elementary particle physics. There is one law of physics that needs to be understood, and that is the relation between energy and speed of a mechanical object. Here we shall discuss this law for the case of objects moving with a speed small compared to the speed of light so that relativistic effects may be ignored. We are talking about that type of energy, kinetic energy, that you may have learned about in high school.[b] The relation between energy and speed is quadratic: if you accelerate a car to a speed of 100 km/h then you need (ignoring friction) four times as much energy (four times the amount of gas) as for accelerating to 50 km/h. Also the converse is true: to bring a car with a speed of 100 km/h to a standstill you need four times as much braking distance as halting a car going at 50 km/h.

> Your gas usage (mileage) will also go up quadratically with the velocity of your car, because the energy going into friction that you must overcome depends quadratically on the velocity.
>
> Quadratic implies approximately doubling for percentage increases. For example, for a vehicle going at 105 km/h compared to a speed of 100 km/h one has a gas usage increase proportional to $105^2 = 105 \times 105 = 11025$, which is approximately a 10% increase compared to $100 \times 100 = 10000$.
>
> The net result is that driving a vehicle with a speed of 5% over some value requires approximately 10% more fuel per km (or mile) as compared to the consumption at the given value.

---

[b]The mathematical equation for kinetic energy: $E = \frac{1}{2}mv^2$.

Furthermore the amount of energy needed is proportional to the mass of the vehicle. To accelerate a car of 2000 kg to some speed you need twice the energy needed to bring a car of 1000 kg to that same speed. That is sort of obvious, because you could see a car of 2000 kg as two cars of 1000 kg tied together.

The considerations above refer to vehicles moving on earth, but they are more generally valid. To bring a car to a speed of 50 km/h on the moon or on Mars would require the same amount of energy as on earth. The mass of a car, element in the calculation, has nothing to do with gravitation. Nonetheless mass is usually measured by means of weighing the object. Since the weight of an object is proportional to its mass that works fine as long as this measurement is always done on the same planet. But if the weight of an object is measured on the moon it will be much lighter than on Earth. Yet its mass, used in the energy calculation, is the same. Thus the measurement of a mass of an object requires the measurement of its weight and in addition there is the conversion factor from weight to mass, different in different gravitational environments.

What is called mass in this book, especially for elementary particles, has in the first instance nothing to do with weight. It is the factor that enters in the calculation if the energy must be computed given the velocity of the object. If you want to have an idea of a mass measuring machine think of the following. Take the object of which the mass is to be measured. Bring it up to some given speed, and shoot it at a plate fixed on a spring. The plate will be pushed in. The amount by which it is pushed in is a measure of the mass of the object. This mass-meter would work equally well on Earth, the moon or Mars, in fact even on a vessel in empty space.

The important thing is that if the mass of a car is known, then the amount of energy needed to bring it to some speed can be calculated. For relativistic speeds the calculation becomes a little bit more complicated, but the principle remains the same. For a given body the energy can be computed from its mass and the velocity by which it moves. That is true for mechanical objects and it is also true for freely moving particles. Conversely, if the energy and velocity of a particle are known then its mass can be computed. Sometimes one knows the speed of a particle and its energy and in this way its mass can be determined. For example, if for a given car it is known how much gas has been used to get to a certain speed it is possible to compute how many people are seated inside that car (provided the mass of the car itself and the average weight of the passengers is known). This is essentially the method by which the mass of a particle with a very small lifetime can be measured. Measure the energy and the velocity and then the mass may be determined.

When the velocity of some material object becomes close to the speed of light things are different from the way described above, and one must take into account Einstein's theory of relativity. In this theory the velocity of light starts playing the role of infinite speed in the old theory. Thus it is not possible to achieve a speed exceeding that of light, and when a material body has a velocity close to the speed of light its energy becomes very large, in fact infinite in the limit of attaining the speed of light. Velocity becomes a poor way of describing the state of motion of an object. In particle physics one almost always works with speeds close to that of light, and a few numbers will make it clear that using velocity becomes very awkward.

A typical cyclotron of the fifties accelerated protons to an energy of 1 GeV (never mind the units at this point). Taking the velocity of light to be 300 000 km/h this implies a velocity of 212 000 km/s for the protons coming out of this machine. In 1960 the first large CERN machine, the PS, accelerated protons to 30 GeV, implying a velocity of 295 000 km/h. The latest CERN machine, LEP, accelerated electrons to an energy of 100 GeV, implying a velocity of 299 999.6 km/s.

A better suited quantity is the momentum. At low speeds momentum and speed are essentially the same (momentum is simply mass times the speed, $p = mv$), and if the speed becomes twice as large so does the momentum. However, at speeds close to the speed of light the relationship changes, and the momentum is very near to the energy divided by the speed of light.

To us the important quantity is the amount of energy (or momentum) a particle carries, not its speed. To obtain the correct relation between energy and momentum one must take the mass-energy (that is the energy corresponding to its mass) into account; even for an object at rest (meaning zero momentum) the energy is not zero, but according to Einstein it is equal to mc^2, where c denotes the speed of light.

For an object with mass m the equation giving the relation between energy E and momentum p for all values of the momentum is:

$$E = c\sqrt{p^2 + m^2c^2} \quad \rightarrow \quad E = mc^2 \quad \text{if} \quad p = 0.$$

The relations between momentum, energy and velocity v are:

$$p = \frac{mv}{\sqrt{1 - v^2/c^2}}, \quad E = \frac{mc^2}{\sqrt{1 - v^2/c^2}}, \quad v = \frac{pc^2}{E}.$$

In units where the speed of light is one the velocity is simply equal to the momentum divided by the energy:

$$v = \frac{p}{E}$$

1.5 Events

In experimental particle physics one is dealing with "events". In your television tube the hit of an electron on the screen is an event. An event is a happening, a reaction between particles. Since particles cannot be controlled in detail events tend to be different from one another. For example, one can never exactly predict where an electron will hit a screen (on your TV screen it will mostly hit within a small square, a pixel, which is precise enough). Also, even in identical initial circumstances there are usually different reaction modes. An unstable particle may sometimes decay into a certain configuration, and sometimes into another configuration. This holds also for radioactive nuclei that can decay into different modes. Usually many events are needed to study a particular reaction. In an interference experiment with light one needs many photons hitting the screen before the pattern can be seen. In particle physics what you see are tracks, memorabilia of an event, registered in some way. From these tracks one must try to reconstruct what happened, and looking at many such cases an understanding can be achieved.

Thus there is no rigorously fixed behaviour for unstable particles. For example, the neutron is an unstable particle that on the average lives for about 10 minutes. It decays into a proton, an electron and a neutrino[c] (in fact, in today's parlance, an antineutrino). However, that does not mean that a neutron will always live 10 minutes. Sometimes it will live 5 minutes, sometimes 10 seconds, sometimes 30 minutes. Only by observing many neutron decay events can one make up an average and determine what physicists call the lifetime (or mean life) of the neutron. It is this probabilistic behaviour that is typical for quantum theory. While it is possible to say something very precise about the average, for any individual event anything is possible.

An important observation can be made here. Looking at neutron decay we see the following: initially there is a neutron. That particle disappears, and three new particles appear, namely a proton, an electron and an antineutrino. Thus in particle reactions particles disappear and new particles are created. This is very important, as this phenomenon, the creation of particles, is at the basis of research at the big accelerators. Particles that are unstable (decay after some time) and that are therefore not present in matter around us, can be recreated in certain reactions. Inciden-

---

[c]The neutrino is a particle not occurring in matter around us. It was discovered while studying neutron decay where it occurs as one of the decay products.

tally, as mentioned before, the neutron contains one up quark and two down quarks. What actually happens is that one of the down quarks decays into an up quark, an electron and an antineutrino The change of a down quark into an up quark transforms a neutron into a proton.

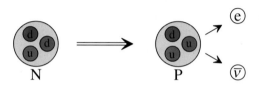

As an interesting aside: neutrons do not age. Thus a neutron that has lived a fraction of a second is indistinguishable from a neutron that has lived 10 minutes. Therefore the probability that a neutron that has lived 10 minutes will decay after another 10 minutes is the same as the probability that the neutron lives 10 minutes. So if the probability for living 10 minutes is 0.5 (that is 50%) then the probability for living 20 minutes is 50% of 50% which is 0.25. And after 30 minutes it is 0.125. In other words, one can figure out the probability of decay or survival for any time, see figure below.

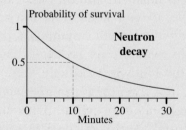

This type of curve is called an exponential curve. It is the same type of curve that you would get for the world population if, say, half the population would disappear every 10 years. Of course the opposite happens, the population doubling about every 15 years (in the developing countries). That is also an exponential curve, one climbing rapidly towards infinity.

One may ask why neutrons in nuclei do not decay. For example, the helium nucleus contains two protons and two neutrons. The point is that this decay is energetically forbidden. If a neutron in a helium nucleus decays then after the decay we are left with a nucleus with one neutron less, and furthermore a proton, an electron and a neutrino. The question is how much energy is needed to remove a neutron from a helium nucleus and replacing it by a proton, because that is what happens (the electron and neutrino just move away, they do not feel any substantial force from either neutron or proton). Replacing a neutron by a proton will generally require extra energy, because the proton is electrically repulsed by the two other protons.

Let us formulate this in a slightly different way. To get a neutron out of a helium nucleus requires energy, you must apply force to pull it out. This energy is called the binding energy. It is usually of the order of a few MeV,[d] although for heavy nuclei (such as for example the uranium nucleus with 146 neutrons and 92 protons) it may be much lower. The binding energy of a proton is usually less than that of a neutron, partly due to the electric force that tends to push the proton away. In other words, it is slightly easier to remove a proton from a nucleus because the electric forces help to push the proton out. Taking that difference in binding energy of proton and neutron into account the decay of a neutron in a nucleus is usually impossible. While the mass of the neutron is larger than the sum of the masses of a proton and an electron, the margin is small (about 0.7 MeV). The difference in binding energies of neutron and proton may (and often will) be more than this small margin and in those cases neutrons in a nucleus cannot decay. There are nuclei for which the energy balance leaves a margin, and then a neutron in such a nucleus can and will decay. That decay is called β radioactivity. Radioactivity was discovered by Becquerel in 1896; subsequent investigations by Pierre and Marie Curie and by Rutherford were crucial in the development of that subject.

---

[d]The MeV is a unit of energy, discussed later in this Chapter.

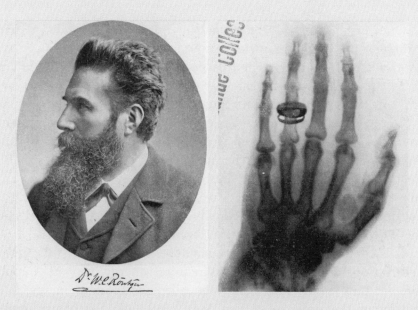

Wilhelm Röntgen (1845–1923). He discovered X-rays. The picture on the right, taken on 22 December 1895, shows what may be the first X-ray picture ever made of a part of a human body. His results, published on 28 December of that year created a sensation around the world, and he demonstrated his invention to Emperor Wilhelm II in Berlin on 13 January 1896. His discovery inspired Becquerel, who thought that the rays had something to do with fluorescence. Investigating fluorescence Becquerel actually discovered radio-activity. In 1899 two Dutch physicists, Haga and Wind, established the wave nature of X-rays. In 1912 the wavelength of X-rays was definitely established by experiments based on a brilliant idea of von Laue (Nobel prize, 1914): they were ultra-ultra-violet light with a wavelength of about 10^{-9} cm (blue light has a wavelength of 4000 Å = 4×10^{-5} cm).

Here is an interview that Röntgen accorded to a journalist:

      J: What did you think?   R: I did not think, I investigated.

      J: What is it?   R: I don't know.

Interestingly, Röntgen was rector of his University (Würtzburg) at the time of his discovery.

It is hard to imagine medicine without X-rays. If there ever was a person that fitted the spirit of Nobel ("who … conferred the greatest benefit to mankind") that was certainly Röntgen. The first physics Nobel prize, in 1901, was awarded to Röntgen. He was a very shy man, and although he gave an excellent banquet speech, he never presented a Nobel lecture. Remember this shy man the next time you enter a hospital!

Marie Curie-Sklodowska (1867–1934) and her husband **Pierre Curie** (1859–1906). Together they discovered that there were elements besides uranium that were radioactive (a term introduced by Marie), notably polonium (named after Marie's country of birth) and radium. Radium, occurring in minute quantities in pitchblende, actually produces heat (nuclear energy): 1 g of radium can heat about 1.3 g of water from the melting point to the boiling point in one hour. The Curies analyzed large amounts of pitchblende, supposedly a worthless residue from mining operations. Pierre died rather young in a traffic incident, which caused intense grief to Marie. Her extended work with radioactive materials was probably part of the cause of her death, 28 years after her husband.

Not only was Marie the first woman to receive the Nobel prize for physics (in 1903, with Becquerel and her husband Pierre), she was also the first person to receive a second Nobel prize (chemistry, 1911: for her services to the advancement of chemistry by the discovery of the elements radium and polonium, by the isolation of radium and the study of the nature and compounds of this remarkable element). The only other woman to receive the physics Nobel prize (1963) is Maria Goeppert-Mayer for her work in nuclear physics. It has remained something of a scandal that Lise Meitner (1878–1968) did not share the 1944 Nobel chemistry prize given to Otto Hahn (1879–1968) for, in fact, their work on nuclear fission (this initiated atomic reactors and nuclear bomb research). It was a clear case of male chauvinism that makes the Swedes blush to this day.

There is also the case of Marietta Blau, see vignette on Powell in Chapter 2.

Incidentally, the reader may have become aware of the fact that there is a force, quite strong, that binds the neutrons and protons inside a nucleus. This force, not always attractive, is much stronger than the electric forces. It has another property, namely it is of limited range. Moving out of the nucleus it becomes rapidly very small. That force, called the strong force, is essentially the same one that binds the quarks in a proton or neutron. The strong force does not affect electrons or neutrinos.

1.6 Electron-Volts and Other Units

A few words on the matter of choice of units of measurement are in order.

As unit of energy we use the eV (electron-Volt) and the related units MeV (Mega-electron-Volt, 1 MeV = 1000 000 eV), GeV (Giga-electron-Volt, 1 GeV = 1000 MeV) and TeV (Tera-electron-Volt, 1 TeV = 1000 GeV). One eV is the energy that an electron gains when passing though an electric field with a potential difference of 1 Volt. The electrons that hit the screen in your TV have an energy of a few thousand eV (a few keV, kilo-electron-Volt), because that many Volts are used for the electric fields that accelerate the electrons inside the tube.

The electron-Volt is a very small energy unit: 1 eV = 1.602 × 10^{-19} joule. One joule is 10^7 erg = 1 watt-s = 1/3600 000 kWh.

Before World War II particle accelerators could produce particles with an energy of up to 16 MeV, nowadays energies of as much as 1 TeV = 1000 000 MeV are reached, using a circular accelerator with a radius of 1 km. In cosmic rays particles with an energy of up to 10^{21} eV = 10^9 TeV occur. No one knows what kind of accelerator is at work somewhere far away in the universe. If we had to achieve that energy with today's technology we would need an accelerator with a diameter of 1000 million km. That is more than 6 times the distance from the earth to the sun (149.6 million km).

In elementary particle physics one usually encounters particles of high energy, that is having speeds close to that of light. The speed of light, commonly denoted by c, is a constant that pops up regularly. It is very convenient to use it as the unit for velocity measurements. That is somewhat like using the speed of sound as a unit of velocity when dealing with fast airplanes; that unit is called the Mach. Here we choose units of length and time such that c equals one.

The speed of light in vacuum is 299792.458 km s^{-1} which is very nearly 300 000 km per second or 3×10^{10} cm s^{-1}. For radio waves or light one may work with wavelengths. The relation between wavelength and frequency is $\lambda = c/v$ or $v = c/\lambda$ where λ is the wavelength and v the frequency. A wavelength of 8000 Å (8×10^{-5} cm, red light) corresponds to a frequency of $\frac{3}{8} \times 10^9$ Mhz. A wavelength of 300 m corresponds to 1 Mhz.

From Einstein we have learned that mass and energy are essentially equivalent ($E = mc^2$), and we can hence use the unit of energy also as a unit of mass. Since we have already taken c equal to one we can express masses directly in eV (or more conveniently, for elementary particle masses, in MeV or GeV).

Another quantity that occurs frequently is Planck's constant, denoted by h. It gives the energy of a photon of light of a given wavelength or frequency. The value of this constant is $h = 4.135669 \times 10^{-21}$ MeV sec. Light or radio waves of a given frequency v is made up from photons that have the energy $E = hv$. This extremely simple and important equation, on a par with Einstein's relation $E = mc^2$, sets the scale for all quantum phenomena.

As an example consider a mobile phone operating at 1 Ghz = 1000 Mhz (megahertz), that is 1 000 000 000 cycles per second. From the antenna of that phone there is a stream of photons each with an energy of $h \times 1 000 000 000 = 4.14 \times 10^{-12}$ MeV. The photons of red light (frequency of about 370 000 Ghz) have an energy of roughly 1.5 eV.

In quantum physics one frequently encounters Planck's constant divided by 2π, where π is what you think it is, namely 3.14.... It is this combination, called the reduced Planck's constant and denoted by $\hbar$, that is usually set to 1.

Setting some constant equal to 1 means choosing units in such a way that that constant becomes one. Thus one chooses the unit of length and the unit of time such that both c and $\hbar$ are 1. These units are called natural units.

> Given the MeV as unit of energy the unit of time is $\hbar$ = 6.582122×10^{-22} s and the unit of length is $\hbar c = 1.97327 \times 10^{-11}$ cm (which is about 1/250 of the size of an atom). A speed of 1 is then equal to $1.97327 \times 10^{-11}/6.582122 \times 10^{-22} \approx 3 \times 10^{10}$ cm/s, the well-known value for the speed of light. If you are considering macroscopic situations then clearly natural units are not very convenient.

Having the speed of light and the reduced Planck's constant equal to 1 greatly simplifies the life of the particle physicist.

1.7 Particle Names and the Greek Alphabet

As more and more new particles were discovered the problem of naming the particles became more and more complicated. In many cases one uses Greek characters; one of the first discovered particles was the muon, denoted by μ, pronounced mu. Also Latin characters are sometimes used to denote particles, for example there is a kaon, indicated by the letter K and there are W's and a Z. Before the muon there was the neutrino, but that name was an Italian invention, derived from the name neutron as both neutrino and neutron did not carry electric charge. The neutrino has a very small or zero mass while the neutron is quite heavy, so you may see the reason. The Italian language has many ways to indicate diminutives: they could have called it neutretto or neutrello or neutrinello. In print the neutrino became quickly designated by means of the Greek letter ν. You will see the names as they come up, but here it may be useful to reproduce the Greek alphabet. Sometimes there are two characters, slightly different, for the same letter. Of course, there are also upper case characters. Even if there is really no one to one relation between the Latin and the Greek characters we have more or less tried to list them in the order suggested by the names.

α	alpha	β	beta	δ	delta	$\epsilon\ \varepsilon$	epsilon
$\phi\ \varphi$	phi	γ	gamma	η	eta	ι	iota
κ	kappa	λ	lambda	μ	mu	ν	nu
ω	omega	o	omicron	$\pi\ \varpi$	pi	$\rho\ \varrho$	rho
$\sigma\ \varsigma$	sigma	τ	tau	υ	upsilon	ξ	xi
ζ	zeta	ψ	psi	$\theta\ \vartheta$	theta	χ	chi

The upper case characters, listed the Greek way:

Γ	Gamma	Δ	Delta	Θ	Theta	Λ	Lambda
Ξ	Xi	Π	Pi	Σ	Sigma	Υ	Upsilon
Φ	Phi	Ψ	Psi	Ω	Omega		

In addition there are a number of upper case characters that are the same as certain Latin characters:

A	Alpha	B	Beta	E	Epsilon	Z	Zeta
H	Eta	I	Iota	K	Kappa	M	Mu
N	Nu	O	Omicron	P	Rho	T	Tau
X	Chi						

1.8 Scientific Notation

Atoms are quite small, the hydrogen atom has a size of about 1 Å. The Ångström, denoted by Å, is one-hundredth of a millionth of a centimeter, or 0.000 000 01 cm. Today the preferred unit is the nanometer, 10 times larger than the Å: 1 nm = 10 Å. When there are that many zeros it is convenient to use the scientific notation: 1 Å = 10^{-8} cm = 10^{-10} m. In scientific notation 2.5 Å could be written as 2.5×10^{-8} cm, which is the same as $0.000\,000\,025$ cm. The basic unit is really the meter (and the derived units cm etc.).

When going to large numbers with many zeros before the decimal points one may use the same scientific notation. For example, 2.5 m $= 2.5 \times 10^{+9}$ nm $= 2\,500\,000\,000$ nm. There are 8 zeros here, not nine, because 2.5 contains already one digit after the decimal point. The + is usually not written, thus $2.5 \times 10^{+9} = 2.5 \times 10^{9}$.

Here is the table for zeros before the decimal point:

deca	hecto	kilo	mega	giga	tera	peta	exa	zetta	yotta
10	10^{2}	10^{3}	10^{6}	10^{9}	10^{12}	10^{15}	10^{18}	10^{21}	10^{24}

Thus 1 kg is 1000 g.

For negative powers with zeros inserted after the decimal point:

10^{-1}	10^{-2}	10^{-3}	10^{-6}	10^{-9}	10^{-12}	10^{-15}	10^{-18}	10^{-21}	10^{-24}
deci	centi	milli	micro	nano	pico	femto	atto	zepto	yocto

2

The Standard Model

2.1 Introduction

In this Chapter we will introduce the known particles and the forces that act between them as we understand today. This ensemble is called the Standard Model. It is a beautiful scheme, with well-defined calculational rules, agreeing well with experiment. It still contains many secrets though, and it may take some time before we will get answers to the questions left open. Even so, the Standard Model represents an enormous body of knowledge of Nature that can be seen as the culmination of 400 years of physics.

Almost everybody has become used to the idea that all matter is a collection of atoms, and that those atoms have nuclei with electrons circling around them. The nuclei are composed of protons and neutrons, and the proton and neutron contain quarks.[a] There is a lot of other stuff going around in the nucleus, but in some rough way this picture contains already much truth. The simplest atom is the hydrogen atom, with only one electron circling a single proton. It occurs in water. Other forms of matter are more complex, but the basic idea is the same: atoms, electrons, nuclei, protons, neutrons, quarks.

---

[a]The name "quark" was introduced by Gell-Mann, from the book *Finnigan's Wake* by James Joyce. He, and independently George Zweig, introduced quarks in 1963. Zweig called them aces and deuces, names that did not stick. For some comments see the book by Robert Serber with Robert Crease, *Peace and War*, p. 200.

Hendrik A. Lorentz (1853–1928) and **Pieter Zeeman** (1865–1943). Lorentz formulated the law of forces exerted by electromagnetic fields on charged particles, in particular on the electron. The experimental physicist Zeeman discovered in 1896 the influence of magnetic fields on light emitted by atoms, and in close collaboration with Lorentz established that this is due to the influence of magnetic fields on the electrons in atoms. They just failed to be the discoverers of the electron: that credit is due to J. J. Thomson. Lorentz and Zeeman shared the second Nobel prize, that of 1902.

Lorentz is also known for his work in the domain of relativity. Prior to Einstein he derived an equation concerning the length contraction of a moving rod. Einstein completed this with his theory of relativity, including time dilatation of moving systems; today the complete set of equations concerning moving bodies is called a Lorentz transformation. Einstein had great respect for Lorentz and expressed that more than once. At the day of Lorentz's funeral all street lamps along the funeral route were draped in black cloth. The telegraph service in the Netherlands was suspended for three minutes at noon. Rutherford and Einstein spoke at the grave.

The idea of a length contraction (although not the actual equation) was also formulated independently by the inventive Irish physicist FitzGerald. After learning about FitzGerald's work, Lorentz, a very scrupulous man, always referred to it.

John J. Thomson (1865–1940). He is generally considered to be the discoverer of the electron, in 1899, when he made a rough determination of the mass of the electron. In those days one measured first the ratio of the charge and the mass of the particle (by studying its motion in a magnetic field), and next the charge. That then allowed a determination of the mass. Lorentz and Zeeman deduced a good value for the charge/mass ratio but they did not measure the charge and also did not use the value for the electron charge quoted in the literature. Thomson received the Nobel prize in 1906.

Thomson measured the electric charge of the electron using a method discovered by his student Charles Wilson (of the cloud chamber). This method relies on the condensation of water vapour around charged particles.

His best theoretical work was done around 1906. He made the important observation that the number of particles in an atom is approximately equal to its atomic weight. Furthermore he noted that the mass of the carriers of positive charge (which is what we now know as protons) is not small compared to the electron mass. Indeed, the proton mass is about 2000 times the electron mass. Thomson was closing in on a model for the atom, but as later papers testified, he got on a wrong track. It took Bohr's genius to clear that up.

As we look at any object, at a table or at our hands, it is curious to realize that all that is but a construction made of particles subject to forces, which from the modern point of view are nothing but the exchange of particles. Particles appear and disappear, and all properties of matter derive from the properties of the constituent particles. From this point of view some properties, often just casually mentioned, turn out to be of overwhelming importance. One of the most striking examples is the difference in mass of two types of quarks, namely of the up and the down quark. These two are the constituents of the proton and the neutron: the proton contains two up quarks and one down quark, the neutron one up quark and two down quarks. Each quark comes in three

Proton   Neutron

varieties, coded red, green and blue, all with precisely the same mass. If in a neutron one down quark is changed to an up quark it becomes a proton. The down quark is more massive than the up quark, and for this reason the down quark can and does decay into an up quark (plus an electron and an antineutrino). Later on in this Chapter we shall introduce other quarks, and quote the masses as experimentally established. There is a certain pattern that you can see in the values of the masses of those quarks. Now the curious thing is this: looking at this pattern, if one had to guess, one would expect that the up quark is more massive than the down quark. However, the down quark is the more massive one and can decay, and therefore the neutron is unstable. One of its down quarks can decay into an up quark and the neutron then becomes a proton (plus some other particles). This small mass difference is of extreme importance for nuclear physics, and therefore for all matter existing. The world would be a very different place if the up quark were more massive than the down quark.

Robert Millikan (1868–1953) and his student **Carl Anderson** (1905–1991). Millikan measured the charge of the electron and delivered the definite experimental proof of Einstein's work on the photoelectric effect. In 1923 he was awarded the Nobel prize. He also was a pioneer in the study of cosmic rays. Anderson is the discoverer of the positron, the antiparticle of the electron, in 1932. Anderson's discovery experimentally vindicated the theoretical idea of antiparticles, proposed by Dirac. Anderson knew vaguely about the Dirac theory, but in his own words "The discovery of the positron was wholly accidental." He was awarded half of the 1936 Nobel prize for this discovery; the other half went to Hess (for the discovery of cosmic rays).

Anderson built a cloud chamber with a strong magnetic field that would curve the tracks of electrically charged particles. He then used this chamber, on the instigation of Millikan, to observe cosmic rays. He discovered that there were "up going electrons", but Millikan told him that "everybody knows that cosmic ray particles go down". What happened was that Anderson initially interpreted positrons as electrons in a magnetic field going in the "opposite direction".

At about the same time, across the ocean, Blackett (Nobel prize 1948) and Occhialini also discovered and correctly interpreted the positron. Anderson, helped by the PR-conscious Millikan, published initially very rapidly in the journal *Science*. His official publication in the *Physical Review* was actually some three months later than Blackett and Occhialini's publication in the *Proceedings of the Royal Society*.

The sun would not shine as that depends on neutron decay. Furthermore the proton would be unstable instead of the neutron. Hydrogen (whose nucleus is a single proton) would not exist as stable matter, and therefore there would be no water! The proton better be stable!

There is a big difference between the mass of the proton and that of the electron. In fact, the proton is about 1800 times heavier than the electron or the positron. The positron is the antiparticle of the electron. It is equally massive but has the opposite charge. Historically it is the first antiparticle observed, by Anderson, in 1932. Energetically it would be easy for a proton to decay into a positron (plus possibly other particles). Luckily for us it does not: there is a special rule followed by Nature that forbids that decay.

2.2 Conservation of Energy and Charge

Some particles are stable, others are unstable. The most important rule here is conservation of energy. In any reaction the final energy must be exactly equal to the initial energy. A particle of a given mass has a certain amount of energy, given precisely by Einstein's equation $E = mc^2$. In asking if a particle can decay, one must first try to find a set of particles whose total mass is less than that of the particle under consideration. A particle with a mass of 100 MeV cannot decay into two particles with a total mass exceeding 100 MeV. The law of conservation of energy forbids this, and Nature is very strict about this law. For more massive particles there will usually be enough energy available, and therefore they tend to be unstable. Excess energy is carried away in the form of kinetic energies of the decay products.

Let us turn once more to neutron decay. The neutron has a mass of 939.57 MeV and it decays into a proton, an electron and an antineutrino:

neutron ⟶ proton + electron + antineutrino

The proton has a mass of 938.27 MeV, the electron 0.511 MeV and the antineutrino mass is very small or zero. One sees that the sum of the masses of the electron and the proton is 938.78 MeV, which is 0.79 MeV less than the neutron mass. From an energy point of view the decay can go, and the excess energy is carried off in the form of kinetic energy of the proton, electron and antineutrino.

However, the energy balance is not the whole story. Why for example is there an antineutrino in this reaction? And why is the proton stable? It could, energy wise, decay into an electron and a neutrino, to name one possibility. Here enters an important concept, namely conservation of electric charge. Charge is always strictly conserved. Since the proton has a charge opposite to that of the electron, that decay, if it were to occur, would have a different charge in the initial state (the proton) as compared with the final state (an electron and an electrically neutral neutrino). Thus there may be conservation laws other than conservation of energy that forbid certain reactions. The law of conservation of charge was already a basic law of electromagnetism even before elementary particles were observed. There are several conservation laws on the level of elementary particles, and some of them remain verifiable macroscopically. Charge and energy are the foremost examples.

On the elementary particle level electric charge has a very special feature: it occurs only in discrete quantities. Measuring the charge in units in which the charge of the electron is -1, one observes charges which are integers, or for quarks multiples of $\frac{1}{3}$. In other words, charge is quantized. This allows us to formulate this conservation law slightly differently; the charge appears as a number, and counting the charge of any configuration amounts to adding the numbers of the various particles. Let us call that the charge number. Conservation of electric charge means that the charge number of the initial state must be equal to that of the final state. For example, for neutron decay (neutron → proton + electron + antineutrino) the charge number of the initial state is zero, while for the outgoing state it is $+1$ (proton) plus -1

Ernest Stückelberg von Breidenbach zu Breidenstein und Melsbach
(1905–1984). This brilliant physicist who introduced the idea of baryon number
(as we call it today) did several things that were Nobel prize worthy; as he
published mostly in a rather inaccessible journal (*Helvetica Physica Acta*), and
moreover not in English, his work went largely unnoticed. He suggested a
finite range for the nuclear forces (Nobel prize to Yukawa, 1949) and he also
developed a formulation of quantum field theory as also done later by
Feynman (see Chapter 9 on particle theory).

Stückelberg suffered from cyclothymie. This leads to manic depressive
periods, and he had to be hospitalized periodically. In his later years he was
always accompanied by a little dog that was claimed to be there to guide him
home in case he lost his way. The dog was always present when his master
gave a seminar, and I have actually witnessed that the dog answered to a
question from the public (in fact, from T. D. Lee) with a short bark while
Stückelberg just watched.

Whenever Stückelberg travelled he took along all of his books and papers
that he might conceivably need. This led to a large number of heavy and big
suitcases and trunks for even the smallest of trips.

In the book by R. Crease and C. Mann, *The Second Creation*, on page 140,
there is a very nice interview with Baron Stückelberg. Memorable is one of his
parting words in that interview: "We live too long."

(electron) which gives zero as well. We may speak of charge as a **quantum number**. The charge quantum number is conserved. This then is our first example of a quantum number: the electric charge of a particle.

2.3 Quantum Numbers

If we were to take the conservation of electric charge as a fact of Nature, then we still do not understand why the proton is stable. It could decay into anything for which the charge would add up to + 1, and for which the combined mass is less than the mass of the proton. There are many possibilities, for example the proton could as far as energy and charge is concerned decay into a positron and one or more neutrinos, or two positrons and one electron. The positron is the antiparticle of the electron, with the same mass but with the opposite charge, that is with positive charge. Why then does the proton not decay into a positron and one or more neutrinos?

In 1938 the Swiss theorist E. Stückelberg did come up with a brilliant idea: perhaps there is another quantum number that must be conserved in all reactions, and perhaps that quantum number would not be conserved for any of the (hypothetical) reactions that would make the proton unstable. Electric charge is quite visible, since it manifests itself directly in the tracks elementary particles make in detectors, but that does not mean that there could not be other quantum numbers that would not be directly visible.

Well, the idea is nice, but how can one verify it? How can one find out about essentially "invisible" quantum numbers? The way it works in general is this: study experimentally many, many particle reactions, and try to catalog which reactions occur and which seem to be forbidden. For example, while the neutron is seen to decay into a proton, an electron and an antineutrino, it does not decay into an electron and a positron, even if that combination has also charge zero and a mass that is only a fraction of the mass of a neutron (1 MeV against 939 MeV). If you have a sufficiently long list, invent a new name and call it a new quantum number. Next try to assign values of this quantum number to the particles in such a way that indeed all absent reactions do violate conservation of this number, while reactions that go do conserve it. One simply tries to systematize the reactions as observed. There is no deep theory, just trial and error.

The above procedure works very well, and in a table of elementary particles one can now find the quantum number assignments that have been found to work. It is pure phenomenology. For example, there is a quantum number called baryon number. Both neutron and proton are assumed to have the value $+1$ for this number. Electron, positron and neutrino have baryon number 0. Therefore, if Nature conserves this quantum number, the neutron cannot decay into an electron and a positron, but it can decay into a proton, electron and antineutrino. Generally, if a particle has some quantum number then its antiparticle must have the opposite quantum number. That has to do with "crossing", a concept that will be discussed in detail in a section further down. Thus antiproton and antineutron have baryon number -1.

In the following we shall encounter a few of these conserved quantum numbers. The one that was discussed above and that makes the proton stable is called "baryon number". The word baryon derives from a Greek word meaning heavy and was introduced by Pais (who also came up with the word lepton). Originally Stückelberg introduced this baryon quantum number to protect the proton from instability. He used the name "heavy charge", and he suggested conservation just like that of electric charge. Later on, systematizing nomenclature, the term baryon number was adopted. The proton and the neutron are assigned the baryon number $+1$, while the photon, electron, positron and neutrino are supposed to have baryon number 0. Conservation of baryon number forbids then decay of the proton into a positron and any number of neutrinos.

It should be emphasized that the stability of the proton is not the only instance where the baryon number conservation law has been observed to hold. It is a law that is generally valid; it prevents proton decay, and that decay is certainly its most stringent test, but it can for example also be seen at work in proton-proton or proton-neutron collisions. In the final state for these processes one must have baryon number $+2$ again. Thus for example two protons, or two neutrons in addition to possibly other stuff. But the reaction

$$\text{proton} + \text{proton} \longrightarrow \text{proton} + \text{proton} + \text{neutron}$$

is forbidden, and is indeed not observed. The initial state here has baryon number $+2$, the final state baryon number $+3$. On the other hand, a reaction such as

$$\text{proton} + \text{proton} \longrightarrow \text{proton} + \text{proton} + \text{neutron} + \overline{\text{neutron}}$$

is allowed. Indeed, the $\overline{\text{neutron}}$, meaning the antineutron, has baryon number -1 and thus the final state has baryon number $+2$, just like the initial state.

2.4 Color

In the table of particles we will encounter a few more quantum numbers, in particular in connection with quarks. There three new quantum numbers pop up, somewhat like electric charge, and the names given are simply the colors red, blue and green. Every quark exist in three varieties: quarks have a green, red or blue charge. There are no color neutral quarks. Thus there exists red, blue and green charge. There exists also negative red charge, and we will call that antired. Similarly for blue and green. Quarks do not have such anticharges, but antiquarks do. Thus there exists three up quarks, with one unit of red, blue or green charge, while the anti-up quark will have minus one unit of red, blue or green charge. We will call a quark with one unit of red charge a red quark, and similarly for the others. An antired antiquark is an antiquark with a value of minus one for the red charge. A red quark and an antired antiquark together make a color neutral combination, much like an electron and positron together are neutral with respect to electric charge. An antired antiquark is simply written as an $\overline{\text{red}}$ quark.

Before going on we must introduce gluons. Gluons are particles of mass zero that interact with the quarks, they are somewhat like photons with respect to electrons. The gluons are responsible for the forces between the quarks, again like the photon is responsible for the electric forces between electrons. Gluons carry color charge, in fact they carry one color and one anti-color charge. For example, there is a red-antiblue gluon. Like photons couple only to charged particles, gluons couple only to colored particles. This will be specified in more detail later on in this section.

There is an important difference between electric charge and color charge. In any reaction, if only one color charge is involved then that color charge is strictly conserved, like electric charge. But if there is more than one color then this is no longer true. As discussed below, three colors may add up to give something that is color neutral.

Macroscopically the color charges are never seen, because quarks never occur singly (in isolation). In other words, bound states of quarks as occurring in stable matter around us are not colored, they are neutral with respect to these color charges. That is like atoms that are electrically neutral. Let us discuss this rather difficult point in some more detail, at the same time trying to make clear why colors have been used to name these charges.

It happens that a very specific combination of equal amounts of red, blue and green may act as color neutral. By this we mean the following.

If there is a bound state of several quarks, then the interaction of any gluon with that bound state is the combination of the couplings of that gluon with the individual quarks. It is now possible to configure a bound state of three quarks of different color, red, blue and green, in such a way that no gluon couples to the combination. That depends critically on the way the quarks are bound together; unfortunately this cannot be explained in a simple manner. The net result, however, can be expressed simply: red, blue and green may combine to something that we may call white, meaning that no gluon couples to it. It is like red, blue and green combining to give white light. Thus in a proton or a neutron one of the three quarks is red, another blue and the third green. Which one is red (or blue or green) cannot be said, they interchange colors all the time. This color changing is effected by means of gluon exchange between these quarks. In order for the combination of these three quarks to be color neutral they must be bound in a very specific way, involving the way the spins of the quarks are oriented inside the proton or neutron.

Consider as an example the hydrogen atom: the nucleus, a proton, carries electric charge and also the electron circling the proton carries electric charge. However, the atom as a whole is electrically neutral, because the electron charge is opposite to that of the proton. Likewise, inside a proton or neutron the three quark colors combine to a neutral color. Seen from a distance, proton and neutron carry no color charge.

We are not saying anything simple here; it is a fact well understood theoretically, but not on an intuitive level. That is of course something that happens all the time in particle physics and the world of quantum mechanics. One can compute many things in great detail, but it is often extremely difficult to "understand" these same things in any easy way. The spooky world of microscopic physics is not at all like our macroscopic world. We are very lucky that the color charges behave very much like ordinary color. Even anti-color makes some sense: take white light and take out the red; what remains is something like antired.

The proton has baryon number 1, and from this one deduces that each of the three quarks in a proton must have baryon number $\frac{1}{3}$. Quarks have color and baryon number. In addition they are electrically charged, quarks occur with charge $+\frac{2}{3}$ or charge $-\frac{1}{3}$, antiquarks with charges $-\frac{2}{3}$ or $+\frac{1}{3}$. A proton contains two quarks

Murray Gell-Mann (1929). He truly dominated particle theory in the sixties. In a systematic way, gradually, he unraveled the immense amount of experimental data on particles that we now understand to be bound states of quarks (Nobel prize 1969). In 1964 he introduced quarks (this was also done, independently, by George Zweig), and like everyone else he was at first quite reluctant to accept them as real particles, as they were never seen singly in any experiment. The situation changed drastically due to experiments at the SLAC machine at Stanford in 1969, influenced strongly by the theoretical work of the particle theorist James Bjorken.

Doing calculations is not Gell-Mann's strongest point. That is probably why he missed out on Cabibbo's theory of quark mixing (see Chapter 3). He certainly knew the basic idea (mentioned in a footnote in a pre-Cabibbo paper with Levy), but did not bring it to fruition. He used to refer to the Cabibbo angle as "that funny angle", which caused Cabibbo to carry the name tag "Funny Cabibbo" at some conference. Earlier, talking about that subject at a Brookhaven conference in 1963 Gell-Mann did not submit his talk for publication, but instead submitted (and indeed published) a page of music of Schubert's unfinished symphony.

Gell-Mann is a passionate bird watcher. That hobby (if you can call it that in this case) relies on extreme honesty in collecting and reporting. I can report that on a trip through Australia he once found himself in a bird aviary near Adelaide; to avoid seeing any bird in captivity he ducked, covered his eyes and rushed through.

with charge $+\frac{2}{3}$ and one quark with charge $-\frac{1}{3}$, resulting in a total charge of $+1$. It turns out that combinations of quarks that are color neutral always have an integral amount of electric charge, never anything like $-\frac{7}{3}$ or $+\frac{5}{3}$.

Theoretically we have some understanding why quark bound states must be color-neutral, and this then explains also why only integral electric charges occur. There is, however, no strict theoretical proof showing that there can be no colored bound states or free particles. This is known under the name of **color confinement**; if there is a color-neutral bound state of several quarks then one cannot take away a single quark, as that would give a colored bound state. The idea is that an infinite amount of energy would be needed to do this separation. The quarks are confined, locked up.

There are yet other quantum numbers, notably electron number to be discussed now.

2.5 The Electron-Neutrino, Electron Number and Crossing

Let us pause for a moment and consider what we have so far. There are the up and down quarks, each in three colors and furthermore the electron and the neutrino. In addition there are the antiparticles corresponding to all these particles. There are other neutrinos to come, and we shall call the one in the decay of the neutron the anti-electron–neutrino. This because it is emitted together with an electron, which turns out to be a general rule. Neutron decay is governed by a quantum number, electron number. Electron and electron–neutrino have electron number $+1$, their antiparticles -1. All other particles have electron number 0. A neutron may thus decay into a proton, electron and an anti-electron–neutrino (and not in proton, electron and electron–neutrino). Thus in this decay the electron number of both the initial and final states is 0.

To understand the significance of electron number we may mention another experimentally observed fact. A neutron decays into a proton, electron and antineutrino. A closely related reaction is a collision type reaction, where a neutrino collides with a neutron. The neutron disappears and one finds as products of this collision a proton and an electron. This reaction is indeed observed (in neutrino experiments). Of course, all neutrinos mentioned here are of the electron type.

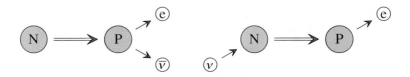

The second reaction, the collision, is precisely what one obtains theoretically when taking the antineutrino from the first reaction (neutron decay) and making it an incoming neutrino. This operation, taking some antiparticle from the final state and turning it into a particle in the incoming state (or vice versa) is called "crossing". Taking a particle from the final state and turning it into an antiparticle in the initial state (and vice versa) is included in this definition. Thus crossing brings us from one process to another.

It is important to note that certain reactions, obtained by crossing, may actually be forbidden by energy considerations. For example the reaction

antiproton $\longrightarrow$ antineutron + electron + anti-electron-neutrino

obtained from neutron decay by crossing both the neutron and the proton, is energetically forbidden, because the antiproton is lighter than the antineutron (they have the same mass as proton and neutron). So it will not occur in reality.

In the collision type reaction shown in the figure above electron number is conserved. Initially there is a neutrino with electron number 1 and in the end there is an electron, also with

Wolfgang Pauli (1900–1958). Pauli introduced the neutrino in 1930. It was not until 1956 that the existence of the neutrino was experimentally proven.

Pauli made many contributions to quantum mechanics and quantum field theory. The best known one is the exclusion principle, stating that no two spin 1/2 particles can be in the same state. This prevents electrons in an atom to crowd all together in the lowest orbit. It is for that discovery that he received the 1945 Nobel prize.

Einstein himself considered Pauli as his successor. Pauli was not aggressive in pushing his own work, but on the other hand he was often very critical about the work of his contemporaries. He discouraged Stückelberg concerning the idea of a particle associated with the strong forces (one that we now call the pion, Nobel prize 1949 to Yukawa). He was equally critical of his own ideas. He wrote down the equations for what we now call the Yang-Mills theory which is the cornerstone of the Standard Model. When he heard Yang talking about it in 1954, he kept asking Yang about some problem arising in those theories, resolved much later through the Higgs particle. There is a lesson here: don't try to solve all problems at once. Also, do not let yourself be discouraged too easily.

During World War II Pauli was at the Institute in Princeton. He was one of the very, very few people who did not want to participate in the atomic bomb project.

There are numerous anecdotes about Pauli. Personally I like the one in which he said, after some seminar, "It is not even wrong."

electron number 1. Neutron and proton have electron number zero. You can see here how the quantum number concept and crossing neatly work together. Essential is that antiparticles have as compared with particles the opposite value for any quantum number, and that crossing also means changing from particle to antiparticle (and vice versa).

Experiments on neutrino reactions similar to the one shown above are done near reactors. These produce enormous amounts of anti-electron–neutrinos. Anti-electron–neutrinos colliding with a proton may produce a neutron and a positron (anti-electron):

$$\text{antineutrino} + \text{proton} \longrightarrow \text{neutron} + \text{positron}$$

That is a reaction where both charge ($+1$ initially and finally) and electron number (-1 initially and finally) are conserved. This is the way that (anti)neutrinos were for the first time explicitly detected by Cowan and Reines, in 1956, near the Savannah River reactor (Nobel prize 1995 to Reines alone, as Cowan died in 1974). Before that date the neutrino was a hypothesis, introduced to explain the missing energy in neutron decay (the difference between the neutron mass and the observed total energy of proton and electron). But now they were seen to do something. That they were actually antineutrinos and not neutrinos was demonstrated by Davis.

The Cowan-Reines reaction is not immediately related by crossing to neutron decay, but rather to antineutron decay. Here is the general rule for any reaction: replacing all particles by their antiparticles gives another possible reaction (called the conjugate reaction). So next to neutron decay there is antineutron decay:

$$\text{antineutron} \longrightarrow \text{antiproton} + \text{positron} + \text{electron–neutrino}$$

Crossing the antineutron, the antiproton and also the electron–neutrino gives the reaction observed by Cowan and Reines.

There is great similarity between a reaction and its conjugated reaction. For example the antineutron mean life is the same as that of the neutron.

2.6 The First Family

The following figure summarizes the particles (except the gluons and photons) mentioned so far. They are the ones that can be found when dissecting matter around us. We speak of the "first family" as there are more families to come.

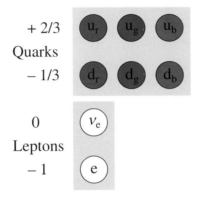

Forces act between these particles, of which electromagnetic interactions are the most familiar. The photon is associated with that. The photon has zero mass and zero electric charge. It interacts with any particle that carries charge, with a strength that increases with the magnitude of the charge. Thus the photon couples stronger to the up quark (charge $+\frac{2}{3}$) than to the down quark (charge $-\frac{1}{3}$), and it does not couple at all to neutrinos or other photons. If two light rays cross they do not scatter each other. All charged particles can emit or absorb photons, but they remain the same particle, for example an electron may become an electron and a photon. This reaction is graphically expressed in the drawing below.

You can impose this figure on any charged particle in the figure of the first family above and that is then a possible reaction. There is no time sequence associated with the figure: the electron can emit or absorb a photon, meaning that the photon can be outgoing or incoming.

To be complete it must be said that quantum effects may induce couplings that originally were not there. Due to that there is, for example, some very weak amount of photon-photon scattering. To understand that requires some understanding of particle theory.

In addition to the particles there exist of course the associated antiparticles. They may be grouped into a figure similar to the one shown above. The antiquarks carry the anti-colors (for example, the anti-color of blue is white minus blue, which is a combination of red and green, which is yellow). Of course, it is very convenient

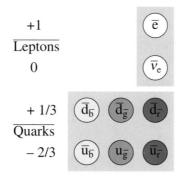

that colors can be used so nicely, but it should be remembered that that is what it is: a lucky accident. Other than that these colors have absolutely nothing to do with the colors of visible light.

The figure for antiparticles is drawn upside down and left-right reflected. Again, the photon may be absorbed or emitted by all antiparticles, with the exception of the electrically neutral anti-neutrino. The same figure as shown before, symbolizing photon interactions, may be used with the antiparticle figure.

The shading in these figures, and the particular way of drawing the antiparticle family has to do with the other known interactions (discussed later), notably the weak interactions of which neutron decay is an example. That decay is due to the decay of a d quark into a u quark (plus electron and antineutrino). In the figure one could represent that by an arrow from the d quark to the u quark. Similarly that same arrow could be used to represent anti-up decay into an antidown quark (plus the same pair, electron and antineutrino). That is why we have drawn the antiparticle family upside down. Since the anti-up quark is lighter than the antidown quark (they have the same masses as the up and down quark) this decay is not actually possible, but reactions obtained by crossing are possible. In fact, crossing changes the second reaction into the first. You could say that the antiparticle figure is the crossed version of the particle figure.

If you feel comfortable at this point brace yourself for the next section, where also particles not present in matter around us are introduced. These new particles are unstable which explains why they are not around us. But they can be produced using accelerators, and that is how we found out about them.

2.7 Families and Forces

The aim of this Chapter is to introduce the elementary particles known today. There are quite a few of them, and there is a very puzzling repetition, not understood at all. What we do here is mainly phenomenology, that is we just shall introduce elementary particles we know to exist and then describe some of their properties. Elementary particles have no further structure that we know of, that is why they are called elementary. Non-elementary particles such as atoms or nuclei or protons and neutrons are bound states of these elementary particles. Also most of the earliest discovered particles such as pions or kaons are bound states. They will be discussed in Chapter 8.

The elementary particles are grouped by one of their fundamental properties, namely spin. The spin of a particle is an internal rotation, much like that of a spinning tennis ball or billiard ball. This spin is quantized, and any given particle has a definite, specific amount of spin. It is measured in a unit whose precise magnitude is not important to us here; the spin can be any

Cecil Powell (1903–1969) and **Donald Perkins** (1925). Powell, Perkins and others had developed photographic methods for studying cosmic rays. Cosmic rays are particles (such as protons) coming to us from the universe around us; they collide with nuclei in the atmosphere and in the collision many particles are produced that could be studied in detail by these methods. In particular this brought clarity concerning particles seen in those collisions. In 1947, Perkins at Imperial College found an event in which a particle (the pion) interacted with a nucleus. Before that one had observed a particle (the muon) that did not interact strongly with nuclei. Thus Perkins was the discoverer of the pion. Somewhat later Occhialini and Powell at Bristol found two events showing decay of a pion into a muon and something else (a neutrino). Perkins found a third event. Clearly, there were two different particles here, and one had to be lighter than the other since else the decay would be impossible. The masses of these two particles were about 135 MeV (pion) and 100 MeV (muon). The pion, a quark bound state, interacts strongly with the protons and neutrons in a nucleus, while the muon does not. Powell received the Nobel prize in 1950. Some believe that Marietta Blau (a woman) should have been included for her work on photographic emulsions.

Perkins received the High-Energy and Particle Physics prize of the European Physical Society in 2001 for his (later) work on neutrino experiments. He studied the scattering of neutrinos from protons and neutrons, notably measuring what are called neutral currents. Also, he measured total cross sections showing evidence for a quark structure.

Giuseppe (Beppo) Occhialini (1907–1993, left) and **Patrick Blackett** (1897–1974). In the twenties Italians made real progress in the research on cosmic rays, and developed coincidence triggers. Occhialini, familiar with these techniques, went to England where he, with Blackett, developed the triggered Wilson cloud chamber. They almost immediately discovered the positron, at about the same time as Anderson. Blackett received the 1948 Nobel prize for the triggered Wilson chamber.

In 1946 Conversi, Piccioni and Pancini discovered the muon in cosmic rays. At the time the existence of the pion had been proposed on theoretical grounds by Yukawa but that particle interacts strongly with nuclei. Conversi *et al.* showed that the particle most seen in cosmic rays, till then assumed to be the pion, did not interact strongly, and they thus established that the particle was not the pion.

At the end of the war Occhialini (who had escaped the Italian fascist regime to Brasil) returned to England, and joined the photographic emulsion group of Powell.

Occhialini was not a lucid speaker, and perhaps that is why he did not share the Nobel prize with Blackett or Powell. Many feel that he should have. He did receive the prestigious 1979 Wolf prize. It should be said that Blackett was always graceful towards Occhialini, more so than Powell. The Nobel lectures of Blackett and Powell testify to that.

Occhialini played an important part in space research, and a satellite that contributed to the discovery of gamma ray bursts was named after him: Bepposax.

multiple of $\frac{1}{2}$ times that unit, including zero. It is not that a given particle is spinning differently at different times: it always spins a definite amount, and only the axis of rotation may be different. Thus a given particle is always spinning at the same rate. You cannot change that. It is a definite property of the particle and it is called its spin. It is perfectly observable, it complicates scattering of particles much like the collision of tennis balls or billiard balls is influenced by their spin.

The particles that we normally associate with matter all have spin $\frac{1}{2}$. The electron as well as the quarks (the quarks make up the protons and neutrons, and thus the atomic nuclei) have spin $\frac{1}{2}$. The particles that we associate with forces (electromagnetic, weak and strong forces) have spin 1, with the exception of the graviton (associated with gravitational forces) that has spin 2. These are facts of life.

Here is a puzzle: experimentally we have never encountered any elementary particle that has spin zero. There is a hypothetical particle, the Higgs boson, that supposedly has spin zero, but this particle has not been observed so far. It plays a very important role in the theory, and it is certainly one of the aims of this book to explain why this particle is hypothesized, and why a massive experimental effort has been initiated to get at it.

Associated with any particle is the corresponding antiparticle. An antiparticle can be defined by the fact that if taken together with the particle one obtains something that has no properties except energy. No charge, no spin, nothing. For example, the antiparticle of an electron is a positron, whose charge is the opposite of the electron. One could say that it has the opposite spin from the electron, since an electron and a positron combined will give as a result something of zero spin if the spin of the electron is opposite to that of the positron. However, as one can change the direction of spin simply by looking at the particle upside down one does usually not consider the direction of spin as one of the quantities describing a particle. But in any case the magnitude of the spin must be the same, and in fact when we speak of the spin of a particle we usually mean the magnitude of its spin. Thus the spin of the positron is the same as that of the electron and if we combine the two, taking the direction of the spins opposite, we may get total spin zero.

The antiparticle of the proton (spin $\frac{1}{2}$) is the antiproton (also spin $\frac{1}{2}$) having negative charge, and a baryon number of -1. The requirement that antiparticles must have quantum numbers opposite to those of the particles puts a strong restriction on the introduction of any quantum number. For example the reaction

$$\text{proton + antiproton} \longrightarrow \text{electron + positron}$$

should be (and is) possible. Both initial and final state have baryon and electron number zero.

As we have noted before, not only elementary particles have antiparticles, but also non-elementary particles, such as the proton, have their anti-companion. They are simply made up from the corresponding antiparticles.

In addition, the mass of an antiparticle is exactly the same as that of the corresponding particle. The positron mass is the same as the electron mass. Theoretically, the existence of antiparticles has been shown to be a consequence of the theory of quantum mechanics combined with Einstein's theory of relativity. It is known under the name CPT theorem. Experiment has verified the validity of this theorem with great precision, notably by comparing masses of particles and antiparticles.

A particle may be equal to its antiparticle. For this to be possible it must be electrically neutral. If it had a non-zero charge its antiparticle would have the opposite charge and thus be different. In fact, it should have no non-zero quantum numbers at all (except spin). An example of such a "self-conjugate" particle is the photon. Another example is the π^0 which is a spinless bound state of a quark and an antiquark.

There is yet one remark to be made. A particle may have its spin aligned with (or opposite to) the direction of motion. The

George Uhlenbeck (1900–1988), **Hendrik Kramers** (1894–1952) and **Samuel Goudsmit** (1902–1978). Uhlenbeck and Goudsmit are credited with the theoretical discovery of the spin of the electron. They did that as graduate students at the University of Leiden. The value of that spin, 1/2, was totally unexpected and possible only within the framework of quantum mechanics. Lorentz and Fermi were very much against. Ehrenfest, their supervisor in Leiden, and also Bohr encouraged them to publish nonetheless. In 1927 Uhlenbeck and Goudsmit joined the physics faculty of the University of Michigan at Ann Arbor, and contributed to the success of the famous Ann Arbor summer symposia.

They were always very graceful with respect to each other. Many felt that they should have received the Nobel prize; Uhlenbeck did receive the Wolf prize for physics in 1979. I happen to know that he gave half of the money to Goudsmit's widow.

Goudsmit led the Alsos mission that had as goal finding out what the Germans and in particular Heisenberg had done about nuclear bomb development during World War II. They dismantled the German reactor at Haigerloch.

Kramers made many contributions to quantum mechanics. His most important one is the idea of renormalization, and the fact that certain anomalies in the spectrum of hydrogen could be expected and calculated. When indeed such an anomaly (the Lamb shift) was observed his ideas were taken up by Feynman, Schwinger and others who then developed the present theory of quantum electrodynamics (see Chapter 9). Kramers was not really recognized publicly until after his death.

figure above shows the idea. In this figure the spin is counter-clockwise, and we speak of a left-handed particle. However, this is a relative statement. If you move along with the particle, and if you go with a speed larger than that of the particle it will from that point of view move in the other direction, i.e. backwards. Then the movement of the spin relative to its motion will be clockwise, and the particle is now right-handed. Thus if there exist left-handed particles then necessarily there exist also right-handed particles, because observers moving with some speed relative to each other should observe the same physics. If something exists for some observer the same thing should exist for any other observer moving with some velocity relative to the first one. It is a slightly abstract point. If the second observer sees a right-handed neutrino then we know that under the appropriate circumstances the first observer could see right-handed ones as well, in some other process. That is the true physical content of the theory of relativity.

However, the above reasoning fails if the particle has no mass and moves with the speed of light. No matter how fast you go after it, it will always move in the same direction with that speed according to the theory of relativity. Thus for a massless particle "handedness" is no more a relative statement. You can have particles that are always left-handed. If it is indeed massless than the neutrino is such a particle. The interactions are such that always a left-handed particle is emitted. It always spins counter-clockwise with respect to its direction of flight, i.e. it is always "left-handed" as shown in the figure. The antineutrino is always right-handed. The spin flips direction when passing from particle to antiparticle.

There is a curious point here. When you collide a neutrino with an antineutrino moving in the opposite direction the spins

point in the same direction. Therefore in that case the spins necessarily add up to spin 1! Conversely, if you see a particle decaying
into a neutrino/antineutrino pair (flying off in opposite directions) then you know that that particle has spin 1. There is actually such a spin 1 particle, called the Z^0. It indeed decays some of
the time into a neutrino–antineutrino pair.

These statements are subject to change if it is found that
neutrinos have mass, and thus do not move at the speed of light.
In that case you could, by going faster than the neutrino, turn a
left-handed neutrino into a right-handed one. Thus if neutrinos do
have mass then there are both left- and right-handed neutrinos.

2.8 The Spin $\frac{1}{2}$ Particles

The spin $\frac{1}{2}$ elementary particles can be divided into quarks and
leptons. The names of elementary particles have come about
historically in a way that is not necessarily relevant today; for
example the leptons (related to the Greek word for "small") were
at one time called that way because the masses of the electron,
the muon and the associated neutrinos are small compared to
the mass of the proton or neutron (called hadrons from the
Greek word for "strong"). Since then we have learned that proton
and neutron contain up and down quarks, and these quarks are
comparatively light. As another example, the mass of the τ lepton
is by no means small, being about twice that of the proton. Yet the
τ and the associated τ–neutrino are called leptons. Today, particles
that are bound states of quarks are often called hadrons.

The figure below shows the known quarks and leptons arranged
in a pattern that clearly displays many of their properties, as
we shall see. The electric charge (the unit of charge is minus the
charge of the electron) is indicated: particles that are on the same
horizontal line have the same charge. There are three "families" or
"generations", groups of six quarks and two leptons, that have
identical properties except for their masses. For example, the top
quark comes in three equal mass varieties, called the top-red,

Burton Richter (1931) and **Samuel Ting** (1936) are credited with the discovery of the charm quark in 1974. Actually, they did not discover that quark, but a bound state of a charm quark and an anticharm quark; the interpretation in terms of a new quark took a few years. Richter and Ting shared the Nobel prize in 1976.

Richter (and his group) did the experiment at SLAC (Stanford Linear Accelerator Center near San Francisco) using electron-positron collisions. Ting (and his group) studied proton collisions at BNL (Brookhaven National Laboratory, Long Island). The discovered quark bound state was called ψ by Richter and J by Ting; today it is known as the J/ψ.

The discovery of the J/ψ was precisely what theory was waiting for. The charm quark was theoretically predicted, but no one had expected a charm-anticharm particle with the properties as measured. It was unstable, but it lived too long. It took some time before it was understood that this was indeed a charm-anticharm bound state, and what precisely the mechanism was. The SLAC people in their unmatched PR skill spoke of the discovery as the "November revolution that turned the wheel". Well, the wheel had already turned a few years before.

CERN failed to discover the J/ψ at the intersecting storage rings where it was produced copiously, and you can understand the tumultuous discussions at CERN after the J/ψ had been discovered. I tried to find out who or what was to blame, but everybody pointed to everybody. Most of the wisdom was after the fact. There was also misery at Frascati as described in Chapter 7.

top-blue and top-green quark, with electric charge $+\frac{2}{3}$, precisely like the up quark that also comes in three varieties, all with electric charge $+\frac{2}{3}$ as well. The mass of the top quark, however, is about 35 000 times that of the up quark. All of this is somewhat bewildering, but that is the way it is.

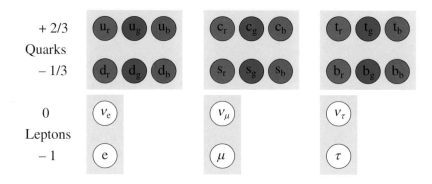

All quarks have baryon number $\frac{1}{3}$ and have color charge, as discussed before. Each quark has one unit of color charge: a red quark has one unit of red charge for example. Color charge can be positive or negative: negative red is called antired. The leptons do not carry color charge. However, they have their own conserved quantum number called lepton number. All leptons shown have lepton number one, the antileptons have lepton number -1. All other particles have lepton number 0. In addition, every lepton pair has its own quantum number. Thus there is electron number (one for electron and electron–neutrino, zero for all other particles) and similarly muon number for the muon and its neutrino and tau number for the tau and its neutrino. Obviously, lepton number conservation is a direct consequence of the conservation of electron, muon and tau number. That may change if neutrinos have masses, because that probably implies a breakdown of the individual leptonic quantum numbers (such as electron or muon number) while not affecting lepton number.

Martin Perl (1927) (left, Nobel prize 1995) is credited with the discovery of the tau particle, in 1975. It is very much like the muon and the electron but much heavier. For example the muon decays part of the time into an electron and a pair of neutrinos, and the tau similarly goes into a muon and a pair of neutrinos. The coupling constants involved are equal within the experimental precision.

The discovery of the tau meant to me personally that there had to be a third family. In this I was way behind: Kobayashi and Maskawa had already argued in 1973 that there should be a third family. Their arguments were based on considerations of quark mixing (discussed in Chapter 3), and at the time they were really hard to swallow although strictly logical.

The Italian physicist **Antonino Zichichi** (1929) was in a sense a forerunner to Perl. He had already been searching for new types of leptons, using antiprotons colliding with protons as well as electron-positron collisions at Frascati. Perl, at Stanford, profited from the higher energy of the positron-electron machine at SLAC.

Zichichi founded and runs a centre for Scientific Culture at Erice, Sicily. It became quite an important part of high energy physics, as summer schools on that subject were organized there yearly.

Perl pushed for a machine that would be dedicated towards tau production. SLAC went instead for another machine, called a B-factory, that would concentrate on the production of particles containing a bottom quark. Currently that appears to have been the right choice.

Here is the greatest puzzle of elementary particle physics today: why are there three families? Are there other families that we have not seen yet? To the latter question we have an answer of which we are reasonably sure: there are no more than three families. The fact that the number of families is fixed makes it more mysterious. Think of the time (1869) that Mendeléev came up with the periodic system of atoms. Today we understand that this comes about as bound states with different numbers of protons and electrons. But here is the problem: bound states normally occur in infinite numbers. You can keep on piling up protons and neutrons to get new nuclei. Eventually they become unstable, but that is another matter. Having only three families and no more makes it virtually impossible to see them as bound states. A further problem is presented by the three neutrinos. For all we know their masses are zero or very nearly so. The difficulty is that no one knows of any way to have a bound state such that the mass of that state is zero. No one understands what is going on. It is very frustrating.

+ 2/3	up 5 MeV	charm 1.3 GeV	top 175 GeV
Quarks			
− 1/3	down 10 MeV	strange 200 MeV	bottom 4.5 GeV
0	el.–neutrino < 0.0000051 MeV	μ–neutrino < 0.27 MeV	τ–neutrino < 31 MeV
Leptons			
− 1	electron 0.511 MeV	muon 105.66 GeV	tau–meson 1777.1 MeV

The figure above shows the names and the masses of the particles. The unit of mass is the MeV or the GeV (1 GeV = 1000 MeV) as described in the section on units.

Not all particle masses are known very precisely. The electron mass is of course quite well measured, it is 0.51099906 MeV with an error of ± 15 in the last two digits. For those who are more familiar with conventional units: in terms of kg this is 0.91093897 divided by 10^{30}. The quark masses, especially the lighter ones, are not so precisely known. For the neutrino masses we have indicated the upper limits. Up to now most people thought that neutrinos are massless, but certain recent experimental facts suggest that neutrinos have (small) masses. If so these masses are less than the limits shown.

There is one more remark to be made. The shaded background indicates a relationship; for example there is some relationship between up and down quarks as concerns the weak interactions. Particles that are not in the same shaded area are not related to each other in any way. So, while we have put the electron and its neutrino in the same family as the up and down quarks, we have no compelling reason for doing so. Perhaps, some day, when we understand the family structure better, we may find that the muon and its neutrino belong in the same group as the up and down quark. The only reason why we have put things as we did is because of mass considerations. We have put the lightest leptons with the lightest quarks.

Here is another major problem of elementary particle physics. Where do all these masses come from? Why is the top-quark so incredibly heavy? Why are neutrinos massless (if they are...)?

It is a sad fact of life that all sophisticated mathematics, all deep considerations that have seen the light of day since 1975 have contributed absolutely nothing towards the three-family problem, nor in fact to a host of other problems that we have not yet talked about. But let us not get ahead; there is a lot that we do understand, and that has been confirmed experimentally.

In addition to these particles there are their antiparticles. They constitute three families, precisely like the ones shown, with the same masses, but with the opposite quantum numbers. Despite the fact that neutrinos are neutral the antineutrinos are still different from the neutrinos: they are not their own antiparticle. They have different handedness as discussed above. Furthermore neutrinos have lepton number 1, and antineutrinos have lepton number -1, which means that some reactions are possible with neutrinos but not with antineutrinos and vice versa.

An antiparticle is usually indicated by drawing a bar above it, and the same holds for color. Thus the anti(red-down-quark) has antired as color. It may combine with the red-down-quark to make something that is color neutral.

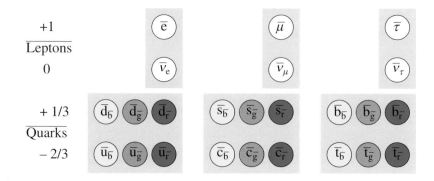

So altogether we now have 18 quarks, 18 antiquarks, 6 leptons and 6 antileptons. The count stands at 48 particles.

2.9 The Spin 1 and 2 Particles

In this section we just enumerate the known particles with spin 1 or spin 2. The following spin 1 particles are known:

Name	Notation	Charge	Mass
vector boson	W^-	-1	80.33 GeV
vector boson	W^+	$+1$	80.33 GeV
vector boson	Z^0	0	91.187 GeV
photon	γ	0	0
gluons (8)	$g_{a\bar{b}}$	0	0

Recall that 1 GeV = 1000 MeV. Compare these masses with the electron mass, about 0.5 MeV and the proton mass, approximately 1 GeV. The vector bosons are really heavy, almost as heavy as 100 protons. The indices a and b for the gluon indicate any of the three colors red, green and blue. Note that there is a bar above the b, which means in fact that the second index indicates any of the three anti-colors, antired, antigreen or antiblue. You might think that there should be 3 × 3 = 9 gluons, but actually there is one absent. It is a "white" gluon having colors that add up to zero (white). It clearly does not exist. There are thus 8 gluons.

The W^+ and W^- are each other's antiparticles. The photon and the Z^0 are their own antiparticles, and the antiparticle of any gluon is simply another one of the gluons. For example, the anti version of the red–antiblue gluon is the blue–antired gluon.

The only known spin 2 particle is the graviton (mass 0); the graviton is to the gravitational field what the photon is to the electromagnetic field. It has not been observed directly, although the gravitational field is of course well-known. The graviton is its own antiparticle.

The particle count is now at 48 + 13 = 61 including the graviton. That's a lot. Our picture of the world is getting complicated again.

2.10 Forces and Interactions

In the macroscopic world two forces are part of our daily life: the electromagnetic and gravitational forces. The reason that these are the only forces that we know by direct experience is because these are long range forces, where long means long compared to the size of a nucleus. Other forces have much shorter ranges. For example, the forces that hold protons and nucleons together in the nucleus are forces with a small range, in practice about 10^{-13} cm. These are basically the same forces that bind the quarks in a proton or a neutron. The weak forces, manifesting themselves in neutron decay, now also observed in many other reactions, notably in neutrino experiments, have a very small range. At low energies (below 20 GeV) they are quite weak, hence the name weak interactions. At higher energies they are about as strong as the electromagnetic interactions.

The concept of a force has grown, historically, from the study of electromagnetic and gravitational interactions. That was a long process, and it evolved from the idea of objects exerting force upon each other into the concept of a field. The latter, due to Faraday, was a major change. The field has an independent existence. It contains energy. To create a field (for example a magnetic field by sending a current through a wire) requires energy. In Newton's time no field was associated with gravitation, no one thought of there existing something in the space between earth and sun. But with electromagnetism it became very difficult not to introduce the concept, given the energy contained in the field. This then led to the idea of an electromagnetic field that could exist and propagate all by itself, as a wave. That in turn led to the idea that light was such a propagating electromagnetic field. It is Maxwell who took that step.

Quantum mechanics made this process even more explicit. Electromagnetic waves consist of photons. So the field idea was replaced by particles. For light that is not that hard to imagine, but what about an electric field around a charged object, for example the electric field around the proton in a hydrogen atom? Is this field also to be seen as a collection of photons?

Indeed, even static fields are seen as collections of photons, although these photons are subtly different from the photons of light. They are "off mass-shell", a concept discussed in Chapter 4. One imagines that the charged source, the proton, continuously emits photons that then move out and later return. This is a very quantum mechanical situation; in the conventional view a photon moving out would be unstoppable and normally not return. In quantum mechanics strange processes like this can happen for short times, longer as the energy of the associated photon is less. An electron passing by the proton might intercept such a photon, absorbing its momentum and energy and thus changing course. This is how we understand scattering of an electron in the electric field of a proton.

In this view the concept of a force does not make any sense. Instead we have interactions, protons or electrons emitting or absorbing photons. What we thought of as a force has become the exchange of a particle. Still, one keeps on talking about forces, so let us go into some detail.

2.11 Classification of Interactions

Interactions may be classified in several ways, and historically this was first done on the basis of their strengths. For example, electromagnetism and gravitation are tremendously different in strength. The gravitational attraction between two protons is down by a factor 10^{36} as compared to the electrical (repulsive) force between those same protons. The only reason that we notice gravitation is because it is collective: the particles in our body feel the sum of the attraction of all particles in the earth. But on the particle level gravitational forces are totally unobservable.

The classification with respect to strength leads to four types of interactions: strong, electromagnetic, weak and gravitational. The photon is central to electromagnetic interactions: all interactions classified as electromagnetic do involve a photon. Similarly strong interactions always involve a gluon, weak interactions almost always the W or Z particles and the gravitational interactions a graviton. In that sense these particles (gluon, photon, W, Z and graviton) are indeed representative for these interactions. The view has become obscured by the fact that the strengths of the interactions are not constant but are energy dependent. At high energy the strong interactions weaken considerably and become roughly equal in strength to electromagnetic interactions. And at low energies the weak forces are so weak that low energy neutrinos have almost no trouble going through the entire earth, while very high energy neutrinos (of the order of 10000 GeV)

Enrico Fermi (1901–1954). In 1934 he published the first theory of weak interactions. He made an analogy between a proton emitting a photon (proton $\rightarrow$ proton + photon) and a neutron emitting an electron-neutrino pair (neutron $\rightarrow$ proton + electron + neutrino). Thus he treated the electron-neutrino pair analogously to a photon. This is in fact quite in line with modern ideas according to which neutron decay essentially goes in two steps: neutron $\rightarrow$ proton + W^- $\rightarrow$ proton + electron + neutrino. In addition to that Fermi was one of the most successful experimental physicists of his era. He directed the construction of the first nuclear reactor and essentially started a whole new chapter of physics by studying pion-proton and pion-neutron collisions.

Fermi was of tremendous importance to US physics as an educator. In 1938 he was told by Bohr that he would get the Nobel prize; since his wife, Laura, was Jewish, they decided not to return from Stockholm to Italy but instead switch to New York, where Fermi became a professor at Columbia University. He later moved to Chicago. Among his students there were Chamberlain, T. D. Lee and Steinberger, to name a few. Thus also through his students did Fermi have a tremendous influence on physics in the US.

Fermi was once asked what Nobel prize winners did have in common. His answer: Not much, not even intelligence.

interact as strongly as charged particles through electromagnetic forces. The classification on the basis of strength alone breaks down. Moreover there are interactions of the same strength as the weak interactions, namely those which we call the "Higgs interactions" that always involve a Higgs particle (a spin 0 particle) and not necessarily a W or Z. Furthermore there are interactions that involve simultaneously a photon and a W or Z or both and possibly a Higgs particle. So, only in a very vague sense can one say that there are electromagnetic forces due to photons and gravitational forces due to gravitons. Indeed we still talk that way, to make contact with the macroscopic reality of those interactions, manifesting themselves as forces. In conclusion we have a large collection of interactions, and all classifications have their limitations.

In this context one meets the concept of a "coupling constant". Such constants are numerical coefficients that occur as a parameter whenever there is an interaction. Generally the strength of an interaction becomes proportional to the magnitude of the associated coupling constant. For example, particles with electric charge interact with electromagnetic fields, thus with photons. This charge functions as a coupling constant. Elementary particles without charge do not interact with photons. Particles with twice the amount of charge interact twice as strongly. And consider gravitation: Newton's gravitational coupling constant is a universal constant that determines the strength of all gravitational interactions. Of course, other factors influence the interaction as well, for example the gravitational interaction is proportional to the masses of the objects.

It is noteworthy that charge appears in two very different ways in particle physics. It appears as a quantum number that is strictly conserved. And it appears as a strength with which particles interact with photons. Here there is a deep theoretical point that we will not explain any further: for the theory to make sense it is for massless spin 1 particles (such as the photon) absolutely essential that the coupling constant be a conserved quantum number. A similar statement can be made about gluons and color charges.

Carlo Rubbia (1934) and **Simon van der Meer** (1925) received the 1984 Nobel prize "for their decisive contributions to the large project, which led to the discovery of the field particles W and Z, communicators of weak interaction". As often, much of this physics progress came from a technological advance, namely the ability to produce a sufficiently dense beam of antiprotons. This was done using a technique called cooling. Antiprotons, originally produced in highly energetic collisions and emerging with more or less random velocities, were deflected, slowed down or accelerated so that they all moved finally at the same pace in the same direction. They were accumulated in a separate storage ring till there were enough of them to produce a sufficiently intensive beam. That antiproton beam was then led into the SPS machine to collide head on with a proton beam, and in the ensuing secondaries enough W's and Z's were produced to allow definite identification.

The protons and antiprotons were thus circulating in the opposite direction in the same machine, the CERN SPS. That machine was originally used to produce 300 GeV protons.

Van der Meer also invented the 'horn of plenty', a focussing device that played an important role in neutrino experiments, extensively discussed in Chapter 7.

Rubbia is not always easy to work with. When he was director of CERN, he changed secretaries at the rate of one every three weeks. This is less than the average survival time of a sailor on a submarine or destroyer in World War II (18 or 6 weeks respectively).

As mentioned above, there is a hypothetical force, namely the Higgs force, involving a Higgs particle. It has not yet been established experimentally.

All in all, the particles that are associated with the various interactions have integral spin, namely zero (Higgs), one (photon, gluon, W and Z) or two (graviton). There could, in principle, exist interactions involving only spin $\frac{1}{2}$ particles although there are theoretical difficulties with such interactions. What we must emphasize is that classification of forces or interactions has become a very tenuous business.

2.12 Electromagnetic, Weak, Strong, Higgs and Gravitational Interactions

For the moment we shall not consider gravitational or Higgs interactions. Studying the remaining three interactions between elementary particles we observe three different strengths, three different coupling constants. The best known one is the electromagnetic coupling constant e. The relevant quantity that always occurs in describing electromagnetic processes is $\alpha_{em} = e^2/4\pi \approx 1/137$.[b] The coupling constant is e, the elementary charge, and α_{em} is the combination that one meets when doing calculations. The transition strength, or the transition probability, which is the quantity observed experimentally, is proportional to the square of the coupling constant.

Next there are the weak interactions. The associated quantity is $\alpha_w = 1/40$. The reader may be curious about the fact that we speak of weak interactions, even with α_w about three times as large as α_{em}. Let us just say here that for certain reasons these interactions are at low energies much weaker than the electromagnetic ones (this has to do with the large masses of the W's and Z). In the early days when the weak interactions were discovered very high energies were not yet available in the

---

[b]This assumes use of the natural system of units, where $\hbar$ and c are one.

laboratories. So in those days these interactions appeared extremely weak (like a million million times weaker) as compared to electromagnetic processes and they were therefore called weak interactions. For example, solar neutrinos have no problem going through the earth. This shows that the neutrino interacts very weakly with matter if the neutrinos are of low energy.

The third type of interactions are the strong interactions. One also speaks of quantum chromodynamics (QCD). These are interactions between colored elementary particles (and their bound states such as neutrons and protons). The associated quantity is called α_{qcd}. It is of the order of 1, but becomes smaller at higher energies.

Let us summarize again these interactions. We start with electromagnetic interactions. These interactions always involve a photon that is either absorbed or emitted. This is our first spin 1 particle. We think that the photon has mass 0, although from an experimental point of view an extremely small mass is still possible (less than 6×10^{-16} eV). The photon couples to any particle with non-zero charge, including the vector bosons to be discussed next.

The weak interactions always involve a so-called vector boson. There are three such bosons, two charged and one neutral. They are denoted by W^+, W^- and Z^0. They are very heavy, 80.33 GeV and 91.186 GeV for the charged and neutral bosons respectively. The W^+ and W^- are each other's antiparticles, the Z^0 is its own antiparticle. The vector bosons couple to each other, and as noted above, the charged vector bosons also couple to the photon.

The strong interactions involve the gluons. There are eight of them, and the interactions are complicated. Each gluon is characterized by a color and an anti-color. The basic interaction is roughly like this. There exists a blue-antired gluon $g_{b\bar{r}}$. When such a gluon hits a red quark it changes that quark into a blue quark. It annihilates the red color and creates the blue color. In this way we have 6 gluons: antired–blue, antired–green, antiblue–red, antiblue–green, antigreen–red, antigreen–blue. Where it gets complicated is when considering the gluons that annihilate the same color as they make. Such as the antiblue–blue gluon. One sometimes calls these gluons "diagonal" gluons. In the first instance there are three of them, but there is one superposition, a mixture, of equal amounts of antired–red, antiblue–blue and antigreen–green that does not exist. That mixture might be called a white gluon, as we understand white as equal amounts of red, blue and green. Hence there are in total 8 gluons. The gluons also couple to each other, except the white gluon (if it existed) that would not couple to the others. The gluons are electrically neutral.

The Higgs interaction is as yet hypothetical. It involves a neutral spin 0 particle called the Higgs particle. The strength of its interaction with any particle is proportional to the mass of that particle, and is very weak (except for the heaviest particles such as the top quark for which its strength actually exceeds that of the weak and electromagnetic interactions).

Finally there is the gravitational interaction. The particle associated with that is called the graviton, and it has spin 2 and zero mass. It has been shown that its mass must be zero. On the level of interacting elementary particles the gravitational interactions are extremely weak, and do not really play any role. The only direct experiments along these lines involve the observation of very slow neutrons, and those do fall down in the gravitational field of the earth like anything else.

2.13 Representing Interactions

It is possible to represent interactions of the various spin 1 particles with the members of the families of spin $\frac{1}{2}$ particles graphically. Let us begin with the photon (denoted by γ). As we all know electrons can emit photons: that is how light and radio waves are made. The latter are made by electrons running up and down in an antenna. Thus an electron can emit a photon. Thus we have the transition:

electron ⟶ electron + photon

or

$$e \longrightarrow e + \gamma$$

The arrow on the line itself shows the direction of the flow of (negative) electric charge. The lower arrow shows the direction the reaction proceeds, i.e. the direction of the flow of time. This reaction can go both ways; when light is absorbed by matter (as in the eye when you look at something) the reaction is

$$e + \gamma \longrightarrow e.$$

We may depict all this as a line going from the electron back to

itself while emitting a photon. We can omit the sense of time here because the reaction can go both ways.

The same transition is possible for any charged particle in the three family figure. So we simply represent a photon interaction by a line emitting a γ. This figure may be attached to any of the

charged particles in the family plot, thus to all except the neutrinos. The same holds also for the anti-family plot, as the antiparticles also couple to the photon. So, this little figure can be placed on any charged particle and also antiparticle and it then depicts a process that actually exists in Nature. Placing this little figure on for instance the anti-τ shows that the anti-τ can emit or absorb a photon. This is then a neat way to show what kind of processes are possible.

A similar procedure may be followed for the vector bosons of weak interactions. For the W^- the basic process is this:

or

$$\text{electron} \longrightarrow \text{el.-neutrino} + W^-$$

$$e \longrightarrow v_e + W^-$$

Again, this same transition may occur between any vertical pair in the three family plot, provided the pair lies entirely within the same shaded area. Thus not between v_e and d_r, for example. This is what we meant earlier when we stated that particles in the same shaded area have some relation to each other; the relation is that they can appear together in an interaction with the W's.

We may represent a W^- interaction by a line connecting the

pair, emitting a W^-. In all cases the two members of the pair differ by one unit of electric charge. This must be so as charge is conserved in these transitions, and the W^- carries off one unit of (negative) charge. For this reason we cannot have a transition from a neutrino to a quark emitting a W^-, or else conservation of charge would be violated. As stated earlier charge is strictly conserved in Nature.

The W^+ can be represented by a similar graph. The basic process is:

or

$$\text{el. neutrino} \longrightarrow \text{electron} + W^+$$

$$v_e \longrightarrow e + W^+$$

This transition may occur within any pair in the shaded regions.

$$\downarrow \nwarrow \!\!\!\!\!\!\searrow W^+$$

The arrow on the W^+ line has been reversed compared to the W^- case; this is to indicate the reversal of the flow of (negative) charge. Later we shall use the arrows on the lines in a slightly different sense, namely to distinguish particles and antiparticles. Since the W^+ is the antiparticle of the W^- our drawing remains correct also with that convention.

There are some complications due to CKM mixing, discussed in Chapter 3. Due to that mixing there is also a transition from an up quark to a strange quark and a W^+. In fact there is a whole set of such family changing interactions, including for example top $\rightarrow$ strange + W^+ and top $\rightarrow$ down + W^+. Here we will not discuss these family-changing interactions.

The Z^0, having no charge, causes transitions much like the photon, from a particle to itself. It can also connect to the neutrinos, unlike the photon. The figure shows the associated graph that can be connected to all particles in the three family plot including neutrinos.

Because we have drawn the antiparticle families upside down the same graphs depicting transitions also apply to the antiparticle plot. For example an anti-electron-neutrino may become a positron by emitting a W^-, and likewise we may have a transition from an antibottom quark (electric charge $\frac{1}{3}$) to an antitop quark ($-\frac{2}{3}$) with the emission of a W^+ (electric charges: $\frac{1}{3} \rightarrow -\frac{2}{3} + 1$).

The strong interactions involve gluons, and the transitions are slightly more complicated. We may have a transition from a red up-quark to a green up-quark if we emit a gluon that carries a red charge and an antigreen charge:

$$\text{up}_{\text{red}} \longrightarrow \text{up}_{\text{green}} + \text{gluon}_{\text{red,antigreen}}$$

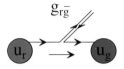

This gluon can do the same for all the other quarks, down, charm etc. We may, as before, represent this gluon without any special reference to the quark type.

We have drawn the gluon as a double line, to show the flow of color charge by means of arrows. This same gluon can also be used on the anti-family drawing, thus may be emitted in case of a transition from an antigreen to an antired quark of any type. It should be emphasized that the arrows on the lines indicate the

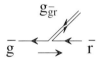

flow of color charge, not the time direction of the transition. Again, at some point arrows on the lines will be used slightly differently, namely to distinguish particles and antiparticles. Colorwise, in the last figure, we have a green charge of -1 becoming a red charge of -1 while emitting a gluon with a green charge of -1 and a red charge of $+1$.

$g_{r\bar{r}}$

r

The "diagonal" gluons couple initially and finally to quarks of the same color. The figure shows how the (red,antired) gluon can be emitted by a transition from a red to a red quark.

The next figure reviews all the particles that we associate with forces. Except for the Z^0, W^+ and W^- they are all massless. There are two diagonal gluons, certain combinations of the antired,red, antiblue,blue and antigreen,green gluons. They are designated by the labels d1 and d2.

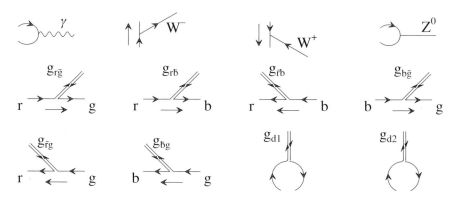

The masses of the photon and all gluons are 0.
The W^+ and W^- masses are both 80.33 GeV.
The Z^0 mass is 91.187 GeV.

There is a particle not shown in the figure: the graviton. It couples to everyone, much like the γ and the Z^0, including the spin 1 particles discussed here with a strength proportional to the particle energy. If the particle is at rest that is essentially the particle mass. However, it also couples to massless particles such as the photon, with a strength depending on the photon energy. A photon passing near the sun on its trip from a star to the earth will be deflected, a phenomenon observable when there is an eclipse (as there happens to be one at the moment of writing this line). This deviation is a result of the graviton-photon coupling.

The hypothetical Higgs boson is not shown either; it couples to all particles with a strength proportional to their masses. If the neutrinos are massless they do not couple to the Higgs boson. But we still have to observe the first Higgs boson! If it exists, its mass must be larger than about 113 GeV.

The figure above is strictly for interactions involving quarks or leptons. We did not include for example the inter-gluon couplings or the photon coupling to the charged W's.

2.14 The Origin of Quantum Numbers

This is perhaps the right place to reconsider the question of quantum numbers. You could say: interactions between particles are always such that quantum numbers are conserved. But this is a question like who was there first, the chicken or the egg. It is in fact of advantage to consider the interactions first and then see what quantum numbers are conserved.

First consider the interactions between quarks and gluons. We observe that in any such interaction at most the color of a quark changes, nothing else. If we count quarks, which is conveniently done using baryon number (every quark has baryon number $\frac{1}{3}$), then evidently this baryon number is conserved. Likewise electric charge is conserved. Since gluons do not couple to leptons nothing there is affected by gluonic interactions.

Considering next the electromagnetic interactions, that is interactions involving a photon, we again see that these interactions involve always the same particle in in- and out-states. For example, the electron emitting a photon: electron $\rightarrow$ electron + photon. Obviously these photonic interactions do conserve just about everything, simply because the same particle occurs initially and finally.

Consider now the weak interactions, i.e. the interactions of vector bosons, either between quarks or between leptons. There are no interactions whereby a quark turns into a lepton. Furthermore, starting with a quark one ends with a quark, although it may be of a different type. Example: up quark $\rightarrow W^+$ + down quark. This then implies again that baryon number is conserved, since all quarks carry the same baryon number $\frac{1}{3}$. Similarly lepton number is conserved. Even more, since these weak interactions on the lepton side are strictly between the lepton pairs of a given family, we do have separate conservation of electron number, muon number and tau number. That may change if neutrinos turn out to have mass.

If there were no CKM mixing then the transitions between the quarks would be strictly between quarks of one and the same family. Thus up $\rightarrow$ strange + W^+ would not occur. Then we would have something similar to electron number etc.; we would have up-down number, strange-charm number and top-bottom number conserved separately. However, there is mixing, and family changing (from up to strange for example) interactions occur. But they occur only in weak interactions, always involving a W^+ or W^-. As these W's are very heavy, interactions at low energy involving these W's are very weak and decay processes are relatively slow. So, there may be quantum numbers that are preserved by all interactions except the weak interactions, and this means that decay processes involving breaking of such a quantum number would be slow. In the old days, before all this was understood, the quantum number "strangeness" was used. As we understand now this amounted to counting the number of strange quarks present. A K-meson, a bound state of quarks involving one strange quark or antiquark, could decay into two pions (no strange quark present) but only weakly. Thus strangeness was partially conserved, as it was respected by all interactions except the weak interactions. Looking at a quantum number by considering interactions we are thus led to the concept of partially conserved quantum numbers.

3

Quantum Mechanics. Mixing

3.1 Introduction

The mechanics of elementary particles is different from that of classical objects such as tennis balls, or planets, or missiles. The movements of these are well described by Newton's laws of motion. The laws describing the motion of elementary particles are given by quantum mechanics. The laws of quantum mechanics are quite different from Newton's laws of motion; yet if a particle is sufficiently heavy the results of quantum mechanics are very close to those of Newtonian mechanics. So in some approximation elementary particles also behave much like classical objects, and for many purposes one may discuss their motion in this way. Nevertheless, there are very significant differences and it is necessary to have some feeling for these.

There are two concepts that must be discussed here. The first is that in quantum mechanics one can never really compute the trajectory of a particle such as one would do for a cannon ball; one must deal instead with probabilities. A trajectory becomes something that a particle may follow with a certain probability. And even that is too much: it is never really possible to follow a particle instant by instant (like one could follow a cannonball as it shoots through the air), all you can do is set it off and try to estimate where it will go to. The place where it will go to cannot be computed precisely; all one can do is compute a probability of where it will go, and then there may be some places where the probability of arrival is the highest. This must be explained, and it

Werner Heisenberg (1901–1976) published his paper introducing quantum mechanics in 1925. The unfamiliar mathematics (matrix calculus) made the paper difficult to read. In 1927 he published his famous uncertainty relations. He made further fundamental contributions to particle physics, for example he recognized that strong interactions are the same for proton and neutron and he found the correct mathematical way to formulate this. He really is one of the all-time greats of physics. In 1932 he received the Nobel prize.

His attitude towards the Nazi regime during World War II may be called ambiguous at best. During the war he was involved in a program aimed at studying uranium fission, but this did not lead to a nuclear bomb. Part of this failure was perhaps due to his poor experimental capabilities for which we may then be thankful.

After World War II Heisenberg was instrumental in the creation of the Max Planck Society with its series of Max Planck institutes. This method of creating centers of excellence has been very fruitful.

In his later years he tried to develop a "theory of everything". It was neither impressive nor successful, and in fact led to rather acerbic comments of Pauli, initially his collaborator.

Erwin Schrödinger (1887–1961) introduced his version of quantum mechanics in 1926. He formulated a wave theory for particles which to this day is the easiest and most often used tool for the quantum mechanical treatment of atoms and molecules. His fundamental equation, the Schrödinger equation, is valid only if the particles involved are not relativistic (speed much less than that of light), which is true for electrons in atoms and atoms in molecules. He received the Nobel prize in 1933.

Schrödinger conceived his ideas during an erotic outburst, spending a holiday in Arosa in Switzerland with an unknown lady. This escapade had apparently an enormous influence on his scientific creativity that for about 12 months remained at a stratospheric level. His life involved many women; his wife Anny maintained a (amorous) relationship with the famous mathematician Hermann Weyl.

The later part of his life, after 1939, was spent at the newly founded Institute for Advanced Studies in Dublin. Remarkably, there appeared to be little problem in this catholic country for him to live there with two women, his wife Anny and Mrs Hilde March (mother of his daughter Ruth).

is done using light as an example, which in pre-quantum physics is described quite accurately by electromagnetic waves. This must be re-examined with the knowledge that light consists of particles, the photons.

The second concept that must be introduced is the idea of an amplitude, a quantity that must be squared to obtain physical statements. That also may be understood by considering light.

3.2 The Two-Slit Experiment

Light, which we know to be nothing but electromagnetic fields, is well described by waves. This was first proposed by Huygens, while in that same period Newton advocated the particle idea. It would have been interesting to go back in time and organize a meeting with these two scientists. One can imagine them looking at a visitor from the future who knows all the answers. Thus, first question by Newton (or Huygens):

What is light: waves or particles?

The answer:

uuuh uuuh .... both.

Probably Newton and Huygens would not be amused; one would have a hard time answering them, which would amount to teaching them quantum mechanics.

If one would want to give an answer that would be a bit more precise one could say the following. The trajectory that a particle is going to follow can approximately be found by doing a calculation with waves. That is what it is, a calculation. It is not true that the particle "is" a wave. It is just that to calculate where it goes one uses wave theory. That is the theory describing its motion. It is not the theory describing the particle itself. The particle remains a small, for all we know point-like, object of definite mass (the mass is zero for light). So the correct answer could have been: light is particles, but their laws of motion are those of waves.

So, light can be described by waves, like also sound is described by waves. Waves can interfere, and the classic experiment to see that is the two-slit experiment. The figure below shows the experimental set-up: a source shines light of a specific color onto a surface containing two openings, two slits. Laser light is excellently suited to this purpose. Further down there is a screen catching the light that passes through the slits. The fact that the light is of a specific color means that the frequency of the light is sharply defined, and hence all photons emitted from the source have the same energy, as given by the Einstein-Planck relation $E = hv$.

To avoid all possible confusion in the argument the source of light is supposed to be so weak that only one photon leaves the source every minute. Thus whenever a photon leaves the source the previous one has since a long time (for a photon) hit the screen. This very slow rate is to make sure that different photons in the beam do not bother each other. It is strictly a single photon experiment.

First one of the slits (call it the first) is kept open, the other is closed.

When the first photon passes through the open slit it will hit the screen somewhere, at a more or less unpredictable place. But sending on photons for hours a pattern develops: the photons will hit the screen in some area that is a widened, blurred image of the slit (the blurring is substantial only if the slit is not too wide). This is understood as diffraction (scattering) of the waves by the edges of the slit. If one knows diffraction theory the image can be computed accurately; the picture on the screen that is built up from individual photon hits will slowly fill out to a picture computed using wave diffraction theory. You may have to wait a few weeks at the rate mentioned, but that is what will happen.

What can we learn from this curious behaviour? First, what is the meaning of the intensity of light for the case of particles? The answer seems obvious: the intensity is proportional to the number of photons. There where the light is intense there are many photons. That is also in line with the idea that the intensity of the light gives the energy density, since a photon has a definite energy. Now the photons are going to make a pattern. There will be many that hit the center, and less towards the edge of the image of the slit. Since the photons arrive one by one there is only one way to interpret this: the pattern on the screen describes the relative probability for the photons to hit the screen at some location. That probability is high where the picture is bright, lower towards the edges. Thus here is the idea: compute what light will do using the theory for the propagation of waves. This gives a pattern, a picture on the screen. That picture represents then the relative probabilities for the photons to hit the screen here or there.

This in a nutshell is quantum mechanics. Since the behaviour of waves is vastly different from classical trajectories of material objects it is not surprising that many have difficulties accepting these ideas. But in the end it is really not that complicated: use waves to compute patterns and that will then give us the probabilities for finding particles here or there.

It is when one tries to explicitly follow how a particle moves, from the source of light, through the slit to the screen that things become difficult. Since it is not the purpose of this book to create difficulties we will not occupy ourselves with questions concerning the whereabouts of the photon on its trajectory from source to screen. It is daydreaming. What counts is what you see on the screen. Do not ask if the particle did follow some continuous path. We do not know about that. Forget about it. For all we care the particle just skips the distance all together and will just hit the screen at some place with a certain probability. We have absolutely no idea if it ever passed through the slit, we never will have, and it cannot be established by any method. The only thing that we can do is compute the probability where it will hit the screen.

What happens if the experiment is repeated with the first slit closed and the other slit open? That is simple: exactly the same pattern will be observed except slightly displaced, because the second slit is slightly displaced relative to the first one.

Now open both slits. The naive person, assuming photons passing through the slits as particles, would say that the new pattern is simply the sum of the two, but that is not the case. There is interference, i.e. there are places where the waves from the first slit cancel out those of the second, and other places where they enhance each other. Using wave theory there is really no problem computing that. In the old days this constituted a convincing proof that Huygens was right and Newton wrong. It just goes to show how careful one must be.

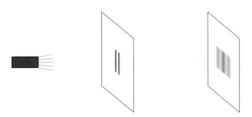

How must this interference be understood? Well, there is nothing special. Compute the pattern to be expected using wave theory and that gives then the probability distribution for the photons such that precisely that pattern comes out in due time. That is the way it is. Individual particles move in unpredictable ways, but in the end, looking at many particles, a pattern forms, of which we can predict the precise form. It is like a roulette wheel: you never know (if you are in a honest place) where the ball will stop, but if you wait long enough it will distribute evenly over all holes. And even if the wheel was loaded there would be a pattern, peaking at some selected places.

Remember now that the experiment was done with the photons strictly separated in time. To say it crudely: they do not interfere with each other, they interfere with themselves. An individual photon moves in a way such that the probability of arrival at some place includes the effect of interference. Of course, the idea of a material particle interfering with itself is quite lunatic, and you will save yourself a lot of headache not trying to visualize that. The interference is in the calculation trying to establish where the photon will go, or rather trying to compute the probability for arriving at a certain place.

3.3   Amplitude and Probability

There is an important consequence to draw from the two-slit experiment. In the calculation one uses waves, coming from both slits and canceling or amplifying each other. Waves may have a sign — think for example of waves on a water surface. Part of a wave is above the average surface (the surface if there was no wave), part is below. When two waves meet there will be interference: the result is that at certain places the water wave will move even more above or below the average surface, while at other places the waves may cancel each other. Now a probability is always positive and not larger than one; a negative probability or a probability larger then one is like saying that you are −20% or more than 100% sure of something. You cannot be less than 0% sure of something. That means already totally unsure. And you cannot be more sure than 100%.

The intensity of the wave is related to the amount the wave goes up or down, either plus or minus. The maximal deviation of the wave from the average (the deviation when on the top or in the valley) is called the amplitude. The intensity is given by the square of the amplitude of the wave, a fact which must now be made plausible.

Consider an idealized situation, where the images of both slits overlap. Then they will enhance each other in the middle,

Max Born (1882–1970). While much less known to the general public than Heisenberg, Dirac or Schrödinger, Born must nonetheless be included as one of the founders of quantum mechanics. He was the one that made the link between the mathematics and physical observation by defining how probability relates to the wave function. That is, he found out that probability is obtained by squaring the amplitude. He received the Nobel prize twenty five years after that work, in 1954.

Born got into discussion with Einstein who refused to accept probability as a fundamental property of physics. It is in a letter to Born that Einstein wrote in 1926 his famous sentence: "Quantum mechanics is very impressive. But an inner voice tells me that it is not yet the real thing. The theory produces a good deal but hardly brings us closer to the secret of the Old One. I am at all events convinced that He does not play dice".

It is amusing to see that Einstein in fact admits that he has no hard arguments against quantum mechanics. He just does not want it. It may have been that he felt that there is something contradictory between quantum mechanics and his theory of gravitation. To this day there is a mystery there, and we do not have a good theory of quantum gravitation. For instance, black holes defy the basic laws of quantum mechanics, and no one has come up with a convincing way to handle that. What to do: disbelieve black holes or quantum mechanics?

interfere out a bit away from the middle, and further on again amplify each other, etc. In the figure above we tried to illustrate that. In the figure below the bold line shows how the intensity varies going through the area horizontally.

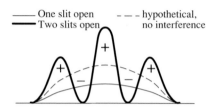

Very, very crudely this is what happens. If only one slit is open there will be some limited area where the light will hit. In the figure the thin line shows the intensity of the light on the screen for this case. If the other slit is open (and the first closed) the same result will be obtained (never mind the slight shift because the slits are slightly displaced with respect to each other). Now have both slits open. There will be light only in the same area as before. However, half the time the waves will compensate, the other half of the time they will enhance. Let us now consider the energy distribution in precise detail. As every photon carries a definite amount of energy that is also the distribution of the photons.

If there is only one slit open the smooth curve drawn with a thin line applies. The maximum amount of energy will be deposited in the centre tapering off towards the sides. The total amount of energy (the total number of photons) is proportional to the surface below that curve.

Open now both slits. If there were no interference then the hypothetical curve (the dashed curve) would apply. The total amount of energy is simply doubled, the surface below the dashed curve is twice the surface below the thin line curve. All that changes is that we get twice as many photons everywhere. In the centre the dashed curve is twice as high as the thin curve. There would be twice as many photons in the centre.

However, there is interference. The total number of photons will still be the same, twice as much as with only one slit open. However, their distribution is changed drastically. In the centre the light waves enhance each other, while slightly off centre they interfere destructively. Photons that (in the hypothetical case) went to the locations slightly off the centre now arrive in the centre. This is indicated by the + and − signs in the drawing. Extra photons in the central region have been taken from the off-centre regions. As a consequence there are twice as many photons in the central area as compared with the no interference case. That is four times as much as with one slit open. In the central region the bold curve is four times higher than the thin curve.

At this point consider the amplitudes of the light waves. There will be a certain amplitude in the centre if there is only one slit open. If there are two slits open, the waves arriving in the centre amplify each other and the result will be a wave with an amplitude twice as large. Think of waves on water. At the top of a wave the water particles are moved upwards by a certain amount. When two equal waves meet, and they are in phase (the tops coincide), the second wave will move the particles up by that same amount, so that all together the wave rises twice as high. Thus, comparing the one slit case with the two-slit case the amplitude in the centre will be twice as large for the two slits case. However, as argued above, the number of photons arriving in the centre is four times as much. What one sees is that if the amplitude doubles the number of photons quadruples. **The number of photons is proportional to the square of the amplitude**. This also cures the problem of a negative sign; even if the amplitude is negative, the probability, related to the square of the amplitude, will be positive.

The total amount of energy deposited on the screen does not change if there is interference. The distribution changes, but whatever there is extra in the centre has been taken away from the neighbouring regions.

A warning here: one must be very careful with arguments of the type given here. Interference is a complex phenomenon.

In quantum mechanics there is in this context a very important point: conservation of probability. The theory must be such that the total probability of a given particle to arrive somewhere on the screen should be 100%. It should go somewhere, and not disappear halfway, and all probabilities should add up to 100% according to the rule that if it does not hit here it must hit somewhere else. If the particle is unstable and if it can decay on its way to the screen then these decay configurations must be included in the total probability count: the probability of arriving at the screen and the probability of decaying somewhere in between should together add up to one.

So here is the result: when there is more than one possible trajectory for a particle there is a wave of some amplitude associated with each of the possibilities. These waves must be superposed (which means addition or subtraction or something in between) producing a wave with another amplitude. The resultant amplitude must be squared and that gives the probability distribution.

Sometimes one reads about machines that create silence. This is the idea: if there is some noisy area (near a highway for example) then set up a speaker system that produces precisely the same sounds but in such a way as to cancel the original sounds. However, remember that energy must go somewhere. If there is somewhere a point where the waves interfere to zero then there is somwhere else another point where they amplify each other. It is not really possible to make a silencing machine. In the end you just add more noise, slightly differently distributed, depending on the wavelength of the sound waves. If you want to cancel out sounds it is necessary to go back to the source and create a situation where then no energy will be released.

In particle physics it is possible to have two amplitudes that cancel each other completely. One must always consider a process as a whole; if two amplitudes cancel completely then nothing can be emitted. For sound there is an explicit example of that, low frequency sound emitted by a loudspeaker not encased in a box. The sound waves emitted by the back of the speaker may go

around and come out front, where they then interfere destructively with the waves produced by the front of the speaker cone.
You will hear nothing. It becomes impossible to pump energy into
the speaker. The cone will flop back and forth without giving off
any substantial amount of energy to the surrounding air. Some air
moves forth and back from the front to the back of the speaker.
Thus some energy is pumped into the movement of the cone itself
and in the movement of the thin layer of air around the speaker,
but it is a minor amount. It is essential that the waves have low
frequency (large wavelength), so that sound coming from the back,
having to travel some distance, remains still out of phase with the
waves from the front. So the effect disappears for wavelengths
smaller than the diameter of the speaker. That is why speakers are
put in boxes: to absorb the low frequency sound produced on the
backside. You can also put the speaker at a hole in a soundproof
wall. That gives quite a good reproduction even of low frequency
sounds on both sides of the wall.

The feature that the energy in a wave is proportional to the
square of the amplitude is quite universal. If the cone of the loudspeaker moves in and out twice as much (compared to some initial
case) the energy emitted is four times as much. This is not an
intuitively appealing result, but that is the way it is. You can easily
confuse yourself by playing in your mind with speakers and imag-
ining what they do. Do not forget that the sound of one of the
speakers may reach the cone of the other speaker and so influence
the movement of that cone. It tends to become complicated.
Speaker technology is a complicated issue. Remember that above
we were talking about monochromatic light. To have sound amplitudes of two speakers sum up the waves must also be monochromatic, that is of the same frequency. And then there is interference
and arguments as given above apply.

Another example can be found in electricity (for those who
know about circuits). If there is a current going through some
circuit then the energy absorbed per second is the wattage,
which is the product of voltage and current. The current itself

Nicola Cabibbo (1935). In 1963 the situation in particle physics was very confusing. There were many particles (now understood as bound states of quarks) that were unstable and decayed in a multitude of modes and strengths.

In a footnote in an article by Gell-Mann and Levy the idea of a fixed ratio between certain decay modes was mentioned. Moreover this was cast in the form of an angle, but no attempt was made to implement this idea. There is actually more to it than just an angle, but never mind.

It was Cabibbo's merit that he succeeded in implementing a complete scheme describing the relative strengths of many decay modes. For example, the angle could be fixed by considering the ratio of pion and Kaon decay ($\rightarrow$ muon + antineutrino). Given then the angle he could precisely compute the decay of the muon ($\rightarrow$ electron + neutrino + antineutrino) from neutron decay. Many people including this author puzzled about these reactions; Cabibbo was at that time working in an office at CERN next to mine and at one point told me that he now understood the relation between neutron and muon decay. He said to me mysteriously, "it is an angle". I said: "Ha ha, I suppose we should call it the Cabibbo angle". The joke was in the end not funny. We now speak indeed of the Cabibbo angle.

It was a revolution that brought order in a very confusing situation, and was of fundamental importance with respect to the further development of particle theory.

is proportional to the voltage, and therefore the energy absorbed is proportional to the square of the voltage (or the square of the current, make your choice). This works actually also for the speakers mentioned. The deviation of the speaker cone is proportional to the current that flows through the speaker coil, and the energy delivered is proportional to the square of the current (the energy is equal to I^2R where I is the current and R is the impedance of the speaker). For a speaker not in a box the impedance is for low frequencies largely inductive and no energy is absorbed by the speaker. On has then a situation analogous to a coil without any cone attached, moving freely in the magnetic field inside the speaker without absorbing any energy. A good speaker system behaves as a pure resistor all through the frequency spectrum.

3.4 Cabibbo and CKM Mixing

Now back to the particle families and their interactions with the three vector bosons, W^-, W^+ and Z^0. There is a small complication, yet with important consequences. First the difference between transition strength and coupling constant, mentioned before, must be emphasized. The coupling constant g involved in the up $\rightarrow$ down $+$ W^+ transition has a certain magnitude. The transition strength, i.e. the transition probability for this reaction, which is what can be observed experimentally, is proportional to α_w which can be obtained by squaring g (and dividing by 4π). In other words, the coupling constant may have a sign (as for example, the electric charge of a particle has a sign), but the transition probability, being proportional to the square of the amplitude and hence to the square of the coupling constant, is of course always positive. In fact, this is basically the same squaring as mentioned in the previous section. The amplitude of the wave corresponding to the particle emitted (the W^+ etc.) is proportional to the coupling constant, and the probability is the square of that. That is not different from the emission of a photon by a charged particle: the electromagnetic field emitted is proportional to the charge of the particle (the coupling constant) and the probability will be proportional to the square of that.

Suppose for the moment that there are only two families, the up-down and charm-strange families. Consider the transitions among the quarks caused by the charged spin 1 particles, W^+ and W^-. These transitions specified above would be strictly a "family business", but the actual situation is different. Earlier it was stated that the up quark can become a down quark, emitting a W^+, and the charmed quark can become a strange quark, emitting a W^+. The negative vector-boson W^- is involved in the opposite transition, like down → up + W^-. The strength of these transitions is the same as among the leptons, like for example neutrino → electron + W^+. In other words, the coupling constant for all these couplings is the same, denoted above by g. This coupling constant universality is an important property that plays a large role in theoretical considerations. The figure below shows the transition amplitudes; they have magnitude L and they are proportional to the universal coupling constant g. The transition probability L^2 is proportional to $\alpha_w = g^2/4\pi$.

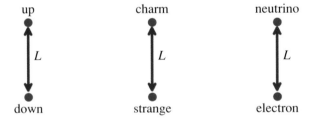

In actual fact the quark transitions are slightly rotated with respect to the family structure. One has that

$$\text{up} \rightarrow \text{down} + W^+$$

goes with a probability slightly less than the lepton transition

$$\text{neutrino} \rightarrow \text{electron} + W^+$$

but the difference equals the probability of a new transition,

$$\text{up} \rightarrow \text{strange} + W^+$$

We ignored energy considerations which actually forbid the reactions as shown. For example, a massless neutrino cannot decay into an electron and a W^+. However, reactions derived from the above by crossing may be possible. Thus the sum of the transition probabilities of the actually observable processes

$$W^- \rightarrow \text{anti-up} + \text{down} \quad \text{and} \quad W^- \rightarrow \text{anti-up} + \text{strange}$$

is equal to the transition probability of

$$W^- \rightarrow \text{antineutrino} + \text{electron}$$

Similarly the sum of the transition probabilities for

$$\text{charm} \rightarrow \text{strange} + W^+ \quad \text{and} \quad \text{charm} \rightarrow \text{down} + W^+$$

is equal to the leptonic transition probability $(v \rightarrow e^- + W^+)$. This whole affair can be viewed as a rotation of the quantity L over an angle, the Cabibbo angle. To explain this consider the figure below, the left part.

The bold line represents the coupling constant for the coupling of the up quark to down and strange quark (plus emission of a W^+). The projection of the bold line on the horizontal axis gives the amount for the down quark coupling, the projection on the vertical axis similarly gives the coupling to the strange quark. As the fat line is horizontal, the coupling to the strange quark is zero.

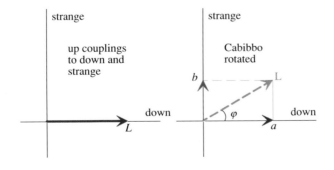

Now rotate the bold line by an angle φ. That rotation is the "Cabibbo rotation". The horizontal projection (indicated by a) is slightly less then in the original figure (where it was equal to L), while there is now a non-zero value for the up to strange transition (b).

A similar situation holds for the coupling of the charmed quark to down and strange quark ($+ W^+$). This is shown in the two figures below. Originally there is no charm to down coupling (the bold line is strictly vertical, no projection on the horizontal axis), after rotation over the same Cabibbo angle there is an amount b for that transition, while the transition to a strange quark is slightly diminished from L to the value a.

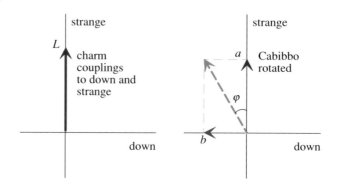

The experimentally determined value for the Cabibbo angle is about 12.7 degrees. The idea of an angle, implying that the probability of some reaction diminishes but that a new reaction takes that up, has been a very fruitful one. At once a lot of poorly understood experimental data started to make sense. In 1963 it was seen that neutron decay (due to the decay $d \rightarrow u$ + electron + antineutrino) proceeded with a coupling constant that was slightly less than that for muon decay ($\mu \rightarrow v_\mu$ + electron + antineutrino). The Cabibbo theory explained that, in perfect agreement with experiment.

Now the question of total probability. It is a property of rotations that the sum of the squares of the components remains the same: the total probability is unchanged. This is a consequence of the well-known theorem of Pythagoras.

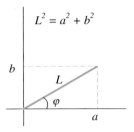

$$L^2 = a^2 + b^2$$

Consider a stick of a certain length. In the figure it is the bold line of length L. From the projections along mutually perpendicular directions the length of the stick can be obtained by using the Pythagorean equation. The sum of the squares of the projections must be calculated, and the length is the square root of that sum. This length L, the length of the stick, is always the same, independent of the angle of rotation, denoted by φ in the picture, and it is directly related to the sum of the squares of the individual components.

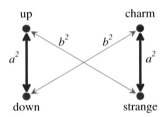

The figure summarizes explicitly the effect of the Cabibbo rotation. Before rotation the transition probability up → down is L^2 (with L equal to that found in muon decay). After the rotation the transition probability of the up quark to go to a down quark is a^2 and to the strange quark b^2, with the sum remaining the same: $a^2 + b^2 = L^2$. Similarly for the charmed quark. The attentive reader may note that in the figure there are arrows on both ends of the lines. This is done to include also the inverse transitions, such as down → up + W^-. It makes the figure inversion invariant, that is, if you turn it upside down it looks the same. The Cabibbo rotation can equally well be discussed considering these inverse reactions.

The rotation may be visualized in a figure, see below. Originally there are two bold lines of equal lengths orthogonal to each other. The Cabibbo rotation rotates these bold lines to the dashed ones. The projections from the dashed line marked with up gives the transitions from the up quark to down and strange quark, and similarly for the charm quark, represented by the dashed line marked charm.

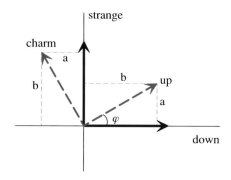

The Cabibbo rotation is experimentally well established, but its origin remains a mystery. The value of the Cabibbo angle, 12.7 degrees, is another number for which we have no explanation, just like for the masses of the various particles. Theoretically there is a relationship to the Higgs particle, but that relationship clarifies nothing. Once more one might hope to understand more if this Higgs particle shows up in the detectors at future machines.

The actual situation is even more complicated because there are three families. There are many more transitions, shown in the figure below. It requires a lot of experimental effort to measure all these transitions and that work is far from completed. Also in this figure we again included the inverse transitions, by attaching arrows to both ends, making the figure invariant under inversion, i.e. turning it upside down.

Makoto Kobayashi (1944) and **Toshihide Maskawa** (1940). In 1973 Kobayashi and Maskawa extended the Cabibbo idea of mixing to three families. At the time there was not even suspicion for the existence of a third family; they did it because in the case of two families they did not have the freedom to accommodate certain data. This concerns the imaginary part of a coupling constant, observed experimentally through the existence of certain reactions. The subject is not discussed here simply because it would require a lot of elaboration.

Anyway, Kobayashi and Maskawa saw that having only two families resulted in a scheme that was too narrow to accommodate all experimental data. In a bold move they assumed the existence of a third family yet to be discovered. In the mood of those days suggesting the existence of a new particle was just "not done". Today many irresponsible people do it. The tau, discovered by Martin Perl in 1975, was the first member of the third family observed experimentally, and gradually the rest of the family was discovered, with at last the top quark being established in 1995.

The story is not finished. A considerable amount of experimental effort is being made to measure and understand that complex coupling constant. At SLAC the B-factory (an accelerator producing lots of bound states of the bottom quark) is at this time running very satisfactorily, giving new information on the subject. The mystery of the complex coupling constant relates to the Higgs particle. Again!

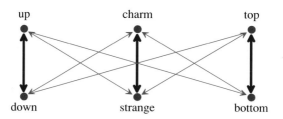

The rotation involves now another axis, the bottom axis. So the figure showing the rotation has become three-dimensional. The next figure is an attempt to visualize this. The bottom axis is assumed out of the paper. The rotation becomes much more complicated: the charm axis moves to the left and slightly forward (out of the paper), and then there is yet another rotation in the up-top plane.

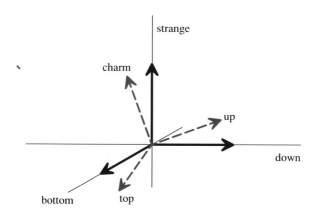

The projections of the bold dashed lines marked up, charm and top onto the third axis (the one sticking out of the paper) give the strengths of the transitions of the up, charm and top quark to the bottom quark. This generalization of the Cabibbo rotation to a rotation of three mutually perpendicular (bold) lines was done by two Japanese physicists, Kobayashi and Maskawa; one hence speaks of the CKM rotation. The remarkable thing is that they did this even without knowing about the third family! They anticipated the existence of the third family on the basis of certain esoteric arguments. We shall discuss some aspects of the CKM rotation. There is no real need to delve into it here, but the facts must be mentioned. You may skip the next two paragraphs.

A rotation in three dimensions is described by three angles: the charm axis is rotated to the left, then rotated forward and finally there is a rotation of the up-top plane, keeping the charm axis fixed. Thus the CKM rotation involves three angles, one of which is the Cabibbo angle. Now here comes something which is truly a matter of quantum mechanics. In quantum mechanics the coupling constants are not just positive or negative, they can be complex, having an imaginary part. If you do not know what complex and imaginary means then that is just too bad, there is really no easy way to explain it. The closest analogy comes from AC electric currents. For an AC current positive or negative is meaningless (except momentarily), but if one considers two currents one can compare them. They can be in the same direction or opposite, but more generally may have a certain phase with respect to each other. Such a phase may be represented by a complex number.

In practice, for the CKM rotation, this means that there is a fourth "angle", and it determines the relative phase of the coupling constants. This angle can be measured in a quite distinct way, it is related to what is called CP violation. But it is outside the scope of this book to explain that in detail.

At this point one may ask if there is also a rotation among the Z^0 couplings analogous to that among the W couplings. In the description given earlier the Z^0 coupling to the down quark for example is

$$\text{down} \rightarrow \text{down} + Z^0$$

One could image that there is mixing between the families, meaning that there could be a coupling such as

$$\text{down} \rightarrow \text{strange} + Z^0$$

However, this is not the case. Such transitions are experimentally seen to be absent to a very high degree, a fact which caused quite some confusion among theorists. Theoretically this is now understood, but explaining that is again outside the scope of this book. Physicists have a way of speaking about the absence of this last reaction: "the absence of neutral strangeness-changing currents". The word neutral refers to the charge of the Z^0. The change of the down quark to the strange quark is referred to as a change of strangeness. The word current refers to the detailed way the Z^0 is coupled to the quarks.

3.5 Neutrino Mixing

One may ask: why is there no mixing among the leptons of the three generations? The answer to that is that we do not know whether there is or not. From the theory it is known that this mixing becomes unobservable if the neutrino masses are all zero. So far the measurements only provide us with upper limits for these masses, and the theory has nothing to say about their possible values. But if the neutrino masses were non-zero there could be something like CKM mixing for neutrinos, and these days a quite large amount of experimental effort is directed towards investigating this issue. Here follows a very simplified discussion.

Consider a solar process involving the emission of a neutrino. That is always due to a transition of the type

$$\text{electron} \rightarrow \text{neutrino} + W^-$$

and crossed versions of this.

Solar neutrino experiments are designed to detect the neutrinos coming from such reactions. If there is no mixing than the neutrino is always an electron-neutrino. If there is mixing, the neutrino emitted in this reaction is some mixture of electron-neutrino, muon-neutrino and tau-neutrino, and that could be observed by considering the reactions induced by these neutrinos. Experimentally that is far from easy, but observations seem to

indicate some mixing. We shall have to await more detailed experimental results. Solar neutrino experiments are among the general
class of experiments over which the experimenter has only limited
control, and for a truly unequivocal proof we will have to wait for
accelerator experiments, of which there are several being built.

Some indication of why masses play a role here may be useful.
Imagine the production of a neutrino as in the reaction mentioned
above. On the detection side precisely the inverse transition is
looked for in the detectors. Since mixing is the same on both sides
you would never know that there is any, one would still obtain an
electron from precisely the mixture emitted primarily.

However, the neutrino must cross some distance from emission
to detection, such as from the sun to the earth. The neutrinos
have a certain energy, and if the masses of the electron-neutrino
and the muon-neutrino (or the tau-neutrino) were different then
they would travel at slightly different velocities. In other words,
the neutrino mixture will change while traveling, and the mixture
observed at detection is no more the same as the one emitted.

The reader may be warned that the above argument is a very
simplified one and should be understood only as an indication why
the distance between emission and detection and the values of the
masses are of importance when observing neutrinos. Quantum
mechanics tells us that the propagation of particles has much to do
with the propagation of waves, and that plays an important role in
these discussions. Even so, there is much truth in the argument.

3.6 Particle Mixing

The strange phenomenon of particle mixing is another exclusively
quantum mechanical effect. Some discussion is in order.

Cabibbo mixing is thought to be the result of a particle mixing
process, so let us take that as an example. Forget for the moment
about the top and bottom quarks. Consider first the case before
rotation. The up quark goes exclusively into the down quark, the
charm quark exclusively into the strange quark.

Abraham Pais (1918–2000). Pais, the author of the books mentioned in the introduction, was a very accomplished physicist. Together with Gell-Mann he published a paper introducing the idea of particle mixing. This was in connection with K^0–$\overline{K}^0$ mixing, a very curious system indeed. When producing a K_0 it would after a while become an $\overline{K}^0$ and the other way around. In the end this resolved itself into a combination of two mixtures, called K_S and K_L. They have very different properties; K_S decays quite quickly, while K_L lives much longer.

Pais introduced the idea of associated production, which is in fact the idea of a new quantum number now called strangeness which had to be conserved in all but weak interactions. Actually, several Japanese physicists published similar ideas at about the same time. This rule explains why certain particles were always produced in pairs (one with strangeness +1, the other –1, so that the sum was 0), given that the initial particles would have no strangeness. This was generally the case, because proton and neutron have strangeness zero, and the new particles were seen in collisions of protons with the protons or neutrons in a nucleus.

Pais, Jewish, living in the Netherlands during World War II, barely survived. He was released from jail just before the end of the war, after an appeal by a very courageous lady armed with a letter from Kramers to Heisenberg (who did not intervene). Perhaps the commanding officer saw the end coming, reason for a leniency extremely rarely seen. A friend of Pais, arrested at the same time, was shot.

Now imagine that there is some special process, some interaction, that causes the strange quark to go over into a down quark, and a similar interaction making the down quark become a strange quark. These things are quite possible, there is nothing that says that particle processes must involve three particles only (such as for example in the process up $\rightarrow$ down + W^+). In fact, one may have transitions involving four particles, or only two particles, and yes, even stranger, only one particle. The latter is really strange, it is like a particle that just stops to exist. Because energy must be conserved that particle must then have zero energy to begin with, but that is sometimes possible. Anyway, let us turn back to the case of two-particle transitions, namely down $\rightarrow$ strange and strange $\rightarrow$ down. Let us suppose that they occur with a certain strength. The reader may ask how it is possible that particles of different mass go over into each other, and indeed that is not possible except for very short times. That will be discussed in Chapter 9, about particle theory. Just do not worry about that aspect now.

Consider now again the process up $\rightarrow$ down + W^+. Since the down quark may now change through this special process into a strange quark we might in the end observe the process up $\rightarrow$ strange + W^+. That would precisely produce the process described through the Cabibbo rotation, and indeed the current philosophy is that this is the mechanism. The situation is slightly more complicated then stated here, because nothing prevents the strange quark from turning into a down quark again and so on. There is a lengthy set of possibilities and it is up to the theorists to figure out what happens in the end. One must consider chains of transitions.

The way these things work out is that there are two very special combinations of the down and strange quark such that they do not change under such a chain of transitions. Let us call these special combinations the Down and Strange quarks. First consider the Down quark, a combination of down and strange quark. What happens is that the down quark in this Down quark can become a strange quark, but on the other hand the strange quark (in this

Down quark) may turn into a down quark. You can imagine that things are such that the net effect is zero, i.e. that there is no change in the total amount of strange quark inside the Down.

Let us give a very crude example. Image a person, Mr A, a dress artist, capable of changing his clothes very quickly. Assume then that he has two sets of clothes, one red, the other green. Suppose further there is a second person, Mr B, capable of the same quick change of dress. He will dress up in whatever is not used by A. If now A changes from red to green, B must give up his green dress and quickly change into the red one.

Assuming that they change clothes quicker than the eye can see what you will observe are two persons with clothes of a color that you can get by mixing red and green. The precise color depends on how Mr A divides up his time in green and red. If Mr A stays, say, for 4 millisec. in red clothes, changes, and stays in green clothes for 2 millisec. etc. he will look some shade of orange. Mr B, staying longer in green, will show a lemon type color. In other words, we will see two persons in a definite complementary color depending on the time distribution of the clothes.

The Down and Strange are the two complementary combinations. In the experiments we will see the Down and the Strange quark, not down and strange. The process whereby two particles turn into certain mixtures because of particle-particle transitions happens just about everywhere where it is possible. An example where no mixing can occur is this: there can never be a transition mixing the up and the down quark. That would involve a change of charge, which nature is careful not ever to do. So, conservation of quantum numbers may prevent certain mixtures. But in general, if two particles have the same quantum numbers (including spin) then they will mix. For example, in principle the up quark could mix with the charm quark, but while that is true it is not observable because the effects of that cannot be distinguished from the effects of down-strange mixing. Cabibbo mixing can be seen as down-strange mixing or up-charm mixing or even a combination of the two, the net result is the same. This is of course why we emphasized earlier the invariance with respect to figure inversion (upside down flipping). In the CKM rotations shown the last figure above you can rotate the bold lines or keep the bold lines fixed but rotate the coordinate system drawn with thin lines. Physicists have opted for the down-strange mixing convention.

Theoretically, the quark mixing described above is thought to be due to the Higgs particle. It may interact with the quarks in a way that produces this mixing. Other interactions never produce the type of particle-particle transitions needed for mixing. This of course is not an explanation, it just shifts the mystery of the CKM rotation to the mystery of the Higgs couplings. When speaking of the theory it is the specific theoretical construction involving this mysterious Higgs particle. It may not be true. So using the word "theoretically" in this Chapter means that it cannot be explained simply, and that it may be wrong.

Another case of mixing concerns the photon and the Z^0. They have the same quantum numbers and they are indeed the final product of some mixing. There is another angle here, called the weak mixing angle, and one speaks of electroweak mixing. The Z^0 couples to the neutrino's, the photon does not as it indeed should not because the neutrino's carry no charge. Here the mixing corrects a potential problem: the photon is precisely that mixture that has no coupling to the neutrino. That is one of the strange effects of mixing: while two particles may both couple to something, it is quite possible that a certain mixture of the two does not. The various possibilities may cancel. Apparently there is a link between electromagnetic and Higgs interactions. A lot of dirt is swept under the Higgs rug!

Theoretically, the Higgs particle is thought to be largely responsible for the CKM mixing, although also other interactions play a role. From the actual mixing as deduced experimentally one may draw important conclusions concerning the Higgs particle and its interactions. The Cabibbo angle can be measured by comparing the processes

up → down + W^+ neutrino → electron + W^+

(and crossed versions). Similarly one may compare the processes

$$\text{up} \rightarrow \text{down} + W^+ \qquad\qquad \text{up} \rightarrow \text{up} + Z^0$$

If there were no mixing they would go at equal strength. By measuring the strength of these transitions the weak mixing angle can be determined.

Earlier some remarks were made concerning gluons of the "diagonal" type. That are gluons whose two colors are each other's anti-color, such as the red/antired gluon. Also here there are mixing possibilities. A red/antired gluon can become a green/antigreen or a blue/antiblue gluon without any violation of quantum numbers. Therefore the actual combinations that propagate are mixtures of these. One of these combinations (one that might be called the "white gluon") is such that in the end it couples to nobody. Since it would never take part in any reaction we may just as well postulate that it does not exist. For the white gluon to play any role would require a new interaction besides the existing quark-gluon couplings.

⌣4⌣

Energy, Momentum and Mass-Shell

4.1 Introduction

The aim of this Chapter is to explain the mechanical properties of elementary particles that will form the basis of much that we shall be discussing. In particular, it is necessary to have a good understanding of momentum and energy, and, for a single particle, the relation between the two, called the mass-shell relation. Energy and momentum are important concepts because of two facts: first they are, in the context of quantum mechanics, enough to describe completely the state of a single free particle (disregarding internal properties such as spin and charge), and second, they are conserved. For energy this is well known: for any observable process the initial energy equals the final energy. It may be distributed differently, or have a different form, but no energy disappears. If we burn wood in a stove the chemical energy locked in the wood changes into heat that warms the space where the stove is burning; eventually this heat dissipates to the outside, but does not disappear. This is the law of conservation of energy. Similarly there is a law for conservation of momentum and we shall try to explain that in this section for simple collision processes.

The fact that a description of the state of a particle in terms of its energy and momentum is a complete description is very much at the heart of quantum mechanics. Normally we specify the state of a particle by its position and its momentum at a given time:

where, when and how does it move. Momentum[a] is a vector, meaning that it has a direction: momentum has thus three components, momentum in the x, y and z directions. That means that for the specification of the state of a particle we have three space coordinates plus the time and the three components of the momentum. In quantum mechanics, when you know precisely the momentum of a particle no information on its location can be given. Heisenberg's famous uncertainty relation forbids this. This relation states that the product of uncertainties in position and momentum is larger than some definite number, thus a smaller uncertainty in one implies larger in the other. There is an analogous relation involving time and energy. Thus knowing the momentum (and thereby the energy) precisely there is nothing more to be known. That's it. If you try to find out where the particle is located at what time, you can with equal probability find it anywhere, anytime in the universe. This is quantum mechanics: to compute the probability of a particle to be somewhere one must use waves; to a particle with a definite momentum corresponds a plane wave, one which extends uniformly over all space. A plane wave is like the waves you see on a relatively quiet sea, extending to the horizon and beyond. This is a very strange subject. It is a difficult subject, because it is a situation very different from daily experience. It is easy to "explain" something that everybody can actually see in the macroscopic world that we live in, but particles do not necessarily behave in that fashion. We must treasure those properties that are the same at the quantum level as well as macroscopically. The laws of conservation of energy and momentum belong to those properties. So this is our way of treating the difficulties of quantum mechanics: talk about things you know and understand, and just do not discuss whatever you cannot know. If you cannot know where the particle is located let us not talk about it.

---

[a]Reminder: at speeds well below the speed of light momentum is simply the product of mass and velocity.

Of course, momentum is never known totally precisely, and one will normally know the location of a particle only in a rough way. With a particle accelerator the particle is part of a beam, and that beam is extracted at some time. So the particle is localized to some extent. But on the microscopic scale these are very, very rough statements, and to deal with a particle exclusively using its energy and momentum is an idealization that for our purposes is close enough to reality. To a particle the beam is the whole universe, and it is big! Here is the scale of things: at a modern accelerator particles are accelerated to, say, 100 GeV, and allowing an uncertainty of 1% in the energy means that as far as quantum mechanics is concerned you can localize it to within one-tenth of the size of a nucleus. An atom is roughly 100,000 times the size of a nucleus, and 100 million atoms make a cm.

So this is what this Chapter is all about: energy and momentum, conservation laws and the mass-shell relation. The discussion will focus on collision processes, the collision of two or more particles. The final state may consist of the same particles but with different momenta and moving in different directions; such processes are called elastic collisions. But it may also happen that the particles in the initial state disappear and other particles appear in the final state. Those are inelastic processes. The conservation laws hold equally well for elastic and inelastic processes, but for inelastic processes there is a difference. The initial particles disappear, and new particles (of which some may be like the initial ones) appear in the final state. Because these particles may have different masses that means that the sum of the initial masses may be different from the total final mass. According to Einstein mass is energy ($E = mc^2$) and therefore this difference in mass implies a difference of energy. That must be taken into account when making up the energy balance. But let us not move ahead of the subject but go about it systematically. Let us state here clearly, to avoid confusion, that when we speak of the mass of a particle we always mean the mass measured when the particle is at rest, not the apparent mass when it moves at high energy.

4.2 Conservation Laws

When thinking about particles most people think of them as small bullet-like objects moving through space. A bit like billiard or snooker balls. There is actually quite a difference between billiard balls and snooker balls: billiard balls are much heavier and do not so much roll as glide over the billiard table while spinning. Billiard balls can be given a spin, which can make their movements quite complicated. As particles generally have spin they resemble billiard balls more than snooker balls. There is another complication in collisions of particles: the particles present after the collision may be very different from those seen initially. A collision of two protons, at sufficiently high energy, may produce a host of other particles, both lighter and heavier than protons. As mentioned before, a collision process where in the final state the same particles occur as initially is called an elastic collision. Billiard ball collisions are elastic collisions, at least if the balls do not break up!

Much of the above picture is correct, although one might do well to think of particles more like blurred objects. Quantum mechanics does that. When particles collide certain conservation laws hold, and some of these laws, valid for particle collisions, also hold for collisions among macroscopic objects as they are made up of those particles. Therefore some of these laws are well-known to us, simply because they can be seen at work in daily life. The foremost conserved quantity is energy: in any collision process the total energy before the collision is equal to the total energy afterwards. Further there is the law of conservation of momentum. There are other conserved quantities (like for example electric charge), but these need not to be discussed in this Chapter. It is good to realize that there may be conservation laws for certain properties that are not at all known on the macroscopic level. One discovers such laws by looking at many, many collision processes and trying to discover some systematics.

Momentum is, for any particle at low speeds, defined as velocity times the mass of that particle. At higher speeds the relationship is more complicated such that the momentum becomes infinite if the velocity approaches the speed of light. Clearly, velocity is not conserved in any process: if you shoot a small object, for example a pea, against a billiard ball at rest then the billiard ball will after the collision move very slowly compared to the pea before the collision. It will however move in the same direction as the pea before the collision. Whatever energy the pea transfers to the billiard ball will have relatively little effect, as that ball is much heavier than the pea. Thus if we are looking for a conservation law the mass must be taken into account, and this is the reason why one considers the product of mass and velocity, i.e. the momentum. So, at the end the pea will be smeared all over the billiard ball, and that ball will have a speed that is the speed of the pea scaled down by the mass ratio, but in the same direction.

Here is a most important observation. **For any object, in particular a particle, momentum and energy are not independent. If the momentum of a particle of given mass is known then its velocity and thus its energy are also known**. This is really the key point of this Chapter. In the following the relationship between momentum and energy will be considered in some more detail. Furthermore, the theory of relativity allows the existence of particles of zero mass but arbitrary energy, and that must be understood, as photons (and perhaps neutrinos) are such massless particles. Also for massless particles the above statement remains true: if the momentum of a particle is known then its energy is known.

Some readers may remember this from their school days: if the velocity of a particle is v then the momentum (called p) of the particle is mv. The kinetic energy is $\frac{1}{2}mv^2$, which in terms of the momentum becomes $\frac{1}{2}p^2/m$. This relation becomes different if the speed is close to the speed of light; relativistic effects start playing a significant role.[b] For a massless particle, always moving at the speed

[b]The precise relation valid for any speed and including the mass energy mc^2 is $E = \sqrt{p^2c^2 + m^2c^4}$.

of light, the energy equals pc, the momentum times c, the speed of light.

It might be noted that velocity has a direction and therefore three components (velocity in the x, y and z direction). Similarly momentum has three components. One says that velocity and momentum are vectors. The relationship between velocity and momentum is vectorial. This means that the relation holds for all components separately, for example the momentum of a particle in the x direction is simply the velocity in the x direction multiplied by the mass of the particle. The conservation of momentum holds for all three components separately, so we have three conservation laws here. The law of conservation of momentum is a law of conservation of a three-dimensional vector. We speak of one conservation law although there are really three separate conservation laws.

Non-relativistically, the total energy (not including the mass energy mc^2) is the sum of the kinetic energies for each of the components of the velocity, thus the total energy equals $\frac{1}{2}mv_x^2$ plus $\frac{1}{2}mv_y^2$ plus $\frac{1}{2}mv_z^2$.

Consider some elastic collision process of two particles called A and B. Think of something like an electron scattering off a proton, or rather from the electric field of the proton. Let us assume that initially one of the particles (B) is at rest while the other (A) moves in with a certain speed, to exit finally at an angle φ (see figure).

In the figure particle B is supposedly much heavier than particle A so that it barely moves after the impact. What precisely the outgoing angle will be depends on the details of the collision. For billiard balls that depends on where precisely the balls hit each other. For elementary particle collisions one never knows positions in any detail, let alone where the particles hit each other.

Accelerators produce beams of a particular type of particle, and such a beam has a size very, very much larger than that of the object that it is aimed at (the nuclei in the target material). Thus one observes many collisions, and a spectrum of angles. Some particles will come out at small angles, some at larger angles etc. How many come out at a given angle depends on the details of the collision, and on the precise way in which the particles interact. In particle physics one thus studies the angular distribution and tries to deduct properties of the interaction. The angular distribution is the distribution of the secondary particles over the directions. For a given time of beam exposure so many particles exit at 10 degrees (for example), so many at 20 degrees, etc.

Such an angular distribution measurement made its entry into physics at Manchester (England), through the historic experiments of Rutherford and his collaborators. A radioactive source emitting alpha particles (these are helium nuclei, containing two protons and two neutrons) was placed in a box with a small hole. The alpha articles going through that hole would pass through a thin foil of gold and subsequently hit a screen. On the screen a picture would evolve, very intense in the centre and tapering off away from that center. See figure.

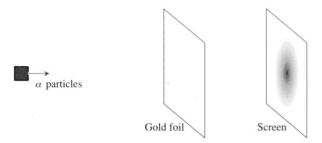

This description and the figure do not do justice to the original experiment: many screens completely surrounding the gold foil were used. Anyway, an angular distribution could be deduced. What was stunning to Rutherford was that some of the α particles actually bounced backwards. That could happen because the nucleus of gold is almost 50 times heavier than an alpha particle. The situation is comparable to a ping-pong ball bouncing back from a billiard ball.

Rutherford himself described his reaction to the back-scattering effect as follows: "It was quite the most incredible event that has ever happened to me in my life. It was almost as incredible as if you fired a 15-inch shell at a piece of tissue paper and it came back and hit you."

At that time one knew Coulomb's law concerning the electric force that two charged particles at some distance exert on each other. This law states that the force becomes much weaker if the distance is larger.

> Precisely, Coulomb's law states that the force is inversely proportional to the distance squared, so if the particles attract (or repulse) each other with a given force at some distance, then the force will be four times weaker if the distance is twice as large. And the force will be four times as large if the distance is halved.

Consider now a heavy, charged particle as target and scatter a much lighter charged particle off it. The deflection strongly depends on the distance at which the light particle is passing the target. For large distance the deviation will be small, but if the particle passes closely by the target it will be deflected strongly. In Rutherford's experiment the alpha particles were not very well collimated and they were evenly distributed over some area much larger than the size of an atom. The angular distribution will reflect the strength of the force depending on the distance.

In 1911 Rutherford, shooting alpha particles at a thin metal foil (for example a gold foil), succeeded in deducing Coulomb's law for the interaction between alpha particles and nuclei (both are electrically charged) from the angular distribution of the scattered alpha particles. To an alpha particle, about 7500 times heavier than an electron, the electrons inside the metal foil are of no importance and it "sees" only the nuclei. On the other hand, in the case of gold, the nucleus is about 50 times heavier than an alpha particle, and it barely recoils under its impact.

This experiment made it clear that atoms are largely empty, with a (heavy) nucleus in the centre. Most of the alpha particles did not seem to collide with anything at all, but some of those that did change direction came out at quite large angles, which is what one expects to happen only if the target is much heavier than the projectile. If a billiard ball hits a light object such as a ping-pong ball it will not deflect substantially, while a ping-pong ball will deflect very much if it hits a billiard ball. Rutherford concluded that the nucleus was very heavy and furthermore that it was at least a hundred thousand times smaller than the atom. It was one of the most important experiments of this era; it opened the door for Bohr's model of the atom, formulated in 1913, after Bohr had spent time at Rutherford's laboratory.

Consider again our example, collision of particle A with particle B at rest. Knowing the mass of particle A the momentum of the initial state can be computed from the initial velocity of that particle (the momentum of particle B is zero, as it is at rest). Let us assume particle A comes out at the angle φ with some particular velocity. Then we can compute the momentum of particle A in the final state. Conservation of momentum will allow us then to deduce the momentum of particle B in the final state: its momentum must be such that combined with the momentum of particle A we get precisely the initial total momentum. Thus if we specify the speed and angle of particle A as it exits, we can compute where B goes from the law of conservation of momentum.

However, there is a complication. We can compute the total energy of the initial state. Since the particles in the final state are the same as those in the initial state we need not to take into account the energy implied by their masses, because that is the same finally as initially. Thus, ignoring the mass-energy (the energy associated with the masses of the particles at rest), the initial energy is just the kinetic energy of particle A. That energy must be equal to the sum of the kinetic energies of the secondary particles. That will generally not be the case for the configuration that we discussed; only for a very specific velocity of the outgoing particle A (at that given angle) will the momentum of particle B be such that the energies of particles A and B add up to precisely the initial energy. Thus conservation of energy has as consequence that in a given direction only one specific momentum is possible.

The figure below shows a configuration with conservation of momentum but without energy conservation. The arrows shown depict the momenta of the particles. Outgoing particles A and B have large momenta pointing roughly in opposite directions. The combination of these two is smaller in magnitude and equal to the initial momentum (the combination of the two momenta is the addition of vectors: one must draw a parallelogram).

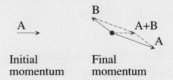

Initial	Final
momentum	momentum

Since the magnitudes of the momenta of A and B are clearly larger than that of A initially (this is depicted by the length of the arrows in the figure), the energies of the final A and B are larger than that of the initial A. Obviously, energy conservation is violated, as the energies of both particles and therefore also their sum exceeds the energy of the initial state.

Briefly, the fact that energy and momentum of a particle are not independent has as consequence that for a given direction for the exiting particle A there is only one specific momentum allowed for that particle, in order for energy to be conserved. Particle A may still exit in all possible directions, but for a given direction the momentum (the speed) is fixed by the law of conservation of energy.

In Rutherford's experiment the mass of the target particle is much larger than that of the incident alpha particle, and in such a case the target particle moves only very slowly after the collision. The energy absorbed by the target is then negligible, and therefore the energy, and consequently the velocity of the outgoing alpha particle is practically the same as before the collision. It just bounces off the nucleus. The alpha particle has four nucleons (two neutrons and two protons), and the target materials used by Rutherford were gold (whose nucleus contains 197 nucleons) and aluminum (27 nucleons). The alpha particle was really the ideal projectile for this experiment: not too light (much heavier than the electrons in the atoms) and not too heavy.

The relation between momentum and energy for a particle of a given mass is a very important relation that will play a central role later on. This relation has a name: it is called the mass-shell relation. It is called that way because of the mathematical figure that one may associate with this relation. Let us make a plot of the energy of a particle versus its momentum.

Momentum is normally a three-dimensional vector, but for the moment we restrict ourselves to a momentum in only one direction. Then we can make a plot, with that single component of momentum along the horizontal axis. We must allow positive and negative values (movement of the particle to the right or the left respectively). We then have for the associated energy a parabola. Remember, the relationship is quadratic: if the energy has some value for a given momentum, then it will be four times greater if the momentum becomes twice as large.

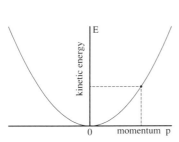

Relationship between momentum and energy of a particle of given mass. Chose some momentum, draw a vertical line (dashed) until it hits the curve and then draw a horizontal line to the energy axis. The energy there corresponds to the momentum chosen

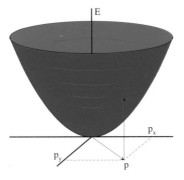

The same for the case of a two–component momentum

If we want to make a plot for the case where the momentum has two components we get a 3-dimensional figure obtained from the previous figure by a rotation around the energy axis. The momentum in the x direction is plotted along the horizontal axis in the plane of the paper, the momentum in the y direction along the axis perpendicular to the paper. That figure looks like a shell, and physicists call it the mass-shell. For a given momentum p, with components p_x and p_y the corresponding energy can be found as shown in the figure. Given p_x and p_y construct the point p. Then draw a line straight upwards from that point p. It will intersect with the shell. The length of the line from p up to the intersection with the shell is the energy E associated with the momentum p.

For relativistic particles, i.e. particles moving with speeds that approach that of light the curve differs slightly from a parabola, as will be discussed in the next section.

4.3 Relativity

If particles have velocities approaching the speed of light relativistic effects become important. The kinetic energy and the momentum depend on the velocity in a different way, namely such that for speeds approaching the speed of light (c) both energy and momentum go to infinity. The speed of light can never be reached, as the energy needed is infinite. That is the way the limit of the speed of light is imposed by the theory of relativity.

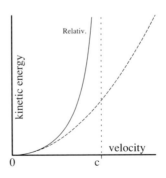

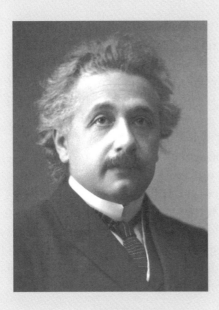

Albert Einstein (1879–1955). The magic year, 1905, when Einstein produced four revolutionary papers (photon, theory of relativity, $E = mc^2$, and an explanation of Brownian motion) was in the period 1902–1908 that he worked at the patent office in Bern. He was actually quite happy there, he liked the work and received reasonable pay. Also his superior was quite happy with him: he was called one of the most esteemed experts at the office. The great advantage of this job was that it left him enough time to do his physics research.

Here are two Einstein anecdotes, of which there are remarkably few.

At some occasion Einstein was received, together with Ehrenfest, by the Dutch queen. As Einstein did not have any formal suit he borrowed one from Ehrenfest; in turn Ehrenfest dug out from his wardrobe some costume that emitted a strong moth-ball odour. This did not go unnoticed by the royalty. As Einstein remarked afterwards: "The royal nose was however not capable of determining which of us two was stinking so badly."

When asked: "What is your nationality?", Einstein answered: "That will be decided only after my death. If my theories are borne out by experiment, the Germans will say that I was a German and the French will say that I was a Jew. If they are not confirmed, the Germans will say that I was a Jew and the French will say that I was a German." In actual fact, Einstein kept his Swiss nationality until his death, in addition to his US citizenship.

In the figure the dashed curve shows the energy versus the velocity in the pre-relativistic theory, the solid curve shows the same
relationship in today's theory. In experimental particle physics one
practically always deals with ultra-relativistic particles, with
speeds within a fraction of a percent (such as 1/100%) from the
speed of light. It is clearly better to work directly with momentum
rather than with velocity.

The relation between energy and momentum changes much less
dramatically when passing from the pre-relativistic formulation to
the relativistic theory. In fact, the energy increases less sharply with
momentum, and for very high values of the momentum the energy
becomes proportional to it (Energy approximately equals momentum times c, the speed of light.) A typical case is shown in the next
figure, with the dashed line showing the non-relativistic case, the
solid curve the relativistically correct relation.

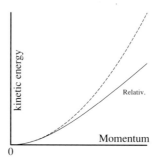

The quantitatively minded reader may be reminded of the equations quoted in Chapter 1. In particular there is the relation between
energy and momentum, plotted in the next figure:

$$E = c\sqrt{p^2 + m^2 c^2}\,.$$

or, using the choice of units such that $c = 1$:

$$E^2 = p^2 + m^2\,.$$

Another important fact is the Einstein equation $E = mc^2$. This very famous equation can be deduced in a number of ways, none of which is intuitively appealing. This equation tells us that even for a particle at rest the energy is not zero, but equal to its mass multiplied with the square of the speed of light. In particle physics this equation is a fact of daily life, because in inelastic processes, where the set of secondary particles is different from the primary one, there is no energy conservation unless one includes these rest-mass energies in the calculation. As the final particles have generally masses different from the primary ones, the mass-energy of the initial state is in general different from that of the final state. In fact, the first example that has already been discussed extensively is neutron decay; this decay is a beautiful and direct demonstration of Einstein's law, $E = mc^2$. Indeed it is in particle physics that some very remarkable aspects of the theory of relativity are most clearly demonstrated, not just the energy-mass equation. Another example is the lifetime of unstable particles, in particular the muon. The lifetime of a moving muon appears to be longer in the laboratory, in accordance with the time dilatation predicted by the theory of relativity.

Thus the mass-energy must be included when considering the relation between energy and momentum. The figure shows the relation between energy and momentum for two different particles, respectively with masses m and M. We have taken M three times as large as m. For zero momentum the energy is simply mc^2 for the particle of mass m and Mc^2 for the particle of mass M.

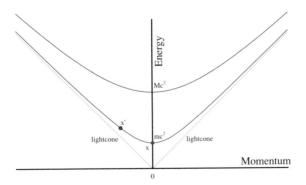

This figure is really the all-important thing in this Chapter. Understanding it well is quite essential, since we shall draw a number of conclusions from it. In itself it is simple: the curve shows the relation between momentum and energy for a single particle. Given the momentum of a particle of mass m one can find the corresponding energy by using the curve for mass m. If the momentum is zero then the energy is mc^2.

In drawing the figure one must make a choice of units. We have drawn a figure corresponding to a choice of units such that the speed of light is one. For very large positive or negative momenta energy becomes very nearly equal to the magnitude of the momentum. In the figure that we have drawn the diagonal lines represent the relation energy = ± momentum. The curves approach these diagonal lines for large momenta. The diagonal lines define the light cone; the reason for that name shall become obvious soon.

To draw the figure we assumed the momentum to have only one non-zero component; if the momentum is in a plane (has two non-zero components) the figure becomes three-dimensional, and can be obtained by rotating the figure shown here around the energy axis. We then have two mass-shells and one cone, the light cone.

One of the results of the theory of relativity is that the velocity of a particle equals the ratio of its momentum and energy (in units where the speed of light is one). So for any point on any one of the curves the velocity is the ratio of the horizontal and vertical coordinates of that point. For large momenta the ratio becomes one (the curve approaches the diagonal line) and the particle moves with a speed very close to one, the speed of light.

In Chapter 1 we gave the relation between momentum, energy and velocity, in particular

$$v = \frac{p}{E}.$$

Here units such that $c = 1$ were assumed. If the particle moves slowly the energy of the particle is very nearly equal to its rest energy, i.e. to $mc^2 = m$. Then $v = p/m$, or $p = mv$.

An interesting point that can be seen from the figure is what happens if we consider the zero mass limit. So, imagine the curve that you get if the point x (fat dot, with mc^2 written beside) is pulled down, to zero. Then obviously the curve becomes the light cone. Thus zero mass particles are perfectly possible, and their energy is equal to the magnitude of the momentum. They always move with the speed of light as the ratio of momentum and energy is always one for these lines. The photon is such a zero mass particle. It has a well defined energy and momentum. Other particles of zero mass are the neutrinos (although there is some question whether their masses are really zero or just very small). Particles definitely of zero mass are the gluons, the basic constituents of the strong forces, and the graviton, responsible for the forces of gravitation.

Finally, the figure may also serve to see what happens to energy and momentum of a particle when changing the reference system from which the particle is observed. First, consider a particle of mass m at rest. The energy will be mc^2, the momentum zero. This is the point x. Now go to a system moving with some velocity v with respect to the particle. In that system the velocity of the particle will be $-v$. The momentum will be what you get by multiplying $-v$ by the mass of the particle. The energy can likewise be computed from this velocity; the momentum and energy are of course related as given by the curve that we have plotted. Thus the new point x' corresponding to the values of energy and momentum in this moving system will be somewhere on the same mass-shell, for example as indicated in the figure. Stated differently: it is impossible to say whether we (the reference system) move or if we are at rest and the particle moves. The relation between momentum and energy is the same. That is in fact precisely the idea of relativity.

4.4 Relativistic Invariance

It was Einstein's theory of relativity that emphasized and made explicit the important role of invariance principles. Already since

Newton and Galilei a number of important laws were generally accepted. For example, there is the idea of rotational invariance: physics in two reference systems that differ from each other by their orientation is the same. Let us formulate this slightly differently. Imagine that two physicists are deducing laws of physics by doing experiments, each in his own laboratory. However, the two laboratories are not quite identical: they are oriented differently, although otherwise there is no difference. For example, imagine that one does his experiments during the day, and when he leaves somebody rotates his whole laboratory over a certain angle, after which the second physicist does his work at night. In the morning the laboratory is rotated back etc. Invariance under rotations means that these two physicists arrive at precisely the same conclusions, the same fundamental laws, the same constants. They measure the same spectral lines when heating up gases, deduce the same laws of electricity (Maxwell's laws), arrive at the same laws of motion etc. Of course, if each of them were to look to the other they would see that they are differently oriented, but it is easy to transform configurations into each other once you know the angles of rotation.

Most people accept this kind of invariance as self-evident. Other examples are translational invariance in both space and time: laws of physics deduced in Europe are the same as those seen in the US, or on the moon. And we also think that the laws of physics are the same today as yesterday or tomorrow. While indeed these invariances do not particularly surprise us, it is only in the twentieth century that we have come to understand their importance. Much of that is due to the fact that things have become much less self-evident with the introduction of the theory of relativity, forcing us to scrutinize these principles more closely. Einstein's theory of relativity was explicitly built upon two principles (in addition to rotational and translational invariance):

— Equivalence of reference systems that are in motion (with constant speed) with respect to each other;

— The speed of light is the same when measured in systems in motion (with constant speed) with respect to each other.

The first principle was already part of physics well before Einstein; it is the second statement that causes effects that are not self-evident. If light is emitted from a moving object one would not expect that light to move with the same speed as a ray coming from an object at rest. You would expect a difference equal to the speed of the moving object. Imagine someone throwing a stone forward while being on a moving train. We would expect that someone standing beside the train would see this stone coming at him with a speed that is the sum of the speed of the stone (as seen on the train) and the speed of the train. Even if the person in the train would merely just drop the stone, the other outside would see that stone coming to him with the speed of the train. Thus the speeds measured on the train or outside the train are not the same. Yet the theory of relativity says that if the stone moves with the speed of light on the train, also the person outside will see it moving with that same speed (we leave aside that it requires infinite energy to get a stone to move with the speed of light). What happened to the speed of the train? Something is funny here. Einstein shifted the problem: a speed measurement implies measurement of distance and time, and these are different from what we normally think, and depend on the state of movement. Thus, there is something funny with time and space measurements. The relationship between a measurement of distance and time of some event by a person on the train to a measurement of the same event by a person outside is very strange to us. In other words, if the person on the train measures the speed of a stone thrown from the train then the speed of that stone measured by the person outside is **not** what you would think, namely the velocity measured on the train plus the speed of the train. To be sure, the deviation is small unless the velocity is in the neighbourhood of the speed of light, so in daily life we see nothing of these effects. The strange thing is the constancy of the speed of light, and that causes all these consequences with respect to measurements

Papers that changed the world: E = mc 2.

Annalen der Physik 20 (1905) 639

13. *Ist die Trägheit eines Körpers von seinem Energieinhalt abhängig?*
von A. Einstein.

Die Resultate einer jüngst in diesen Annalen von mir

~ ~ ~

Die Masse eines Körpers ist ein Maß für dessen Energieinhalt; ändert sich die Energie um L, so ändert sich die Masse in demselben Sinne um $L/9.10^{20}$, wenn die Energie in Erg und die Masse in Grammen gemessen wird.

Bern, September 1905.

In this short (3 pages) paper Einstein presents a derivation of the relation $E = mc^2$. He explicitly gives the equation in words, in the form $m = E/c^2$. The square of the speed of light is given as 9×10^{20}, indeed the square of $c = 3 \times 10^{10}$ cm/s. Energy is denoted by L.

It is interesting to note that he does not present this equation saying how much energy is contained in a given amount of mass. Instead he says: you can measure the energy of a body by measuring its mass. He did not think of mass of a body as a source of energy, rather he saw it as a way of measuring the energy contained in that body. Whether you can get it out is an entirely different matter.

Is the Inertia of a Body Dependent on its Energy Content?

by A. Einstein

The results of a recently published investigation by me in these Annals...

~ ~ ~

The mass of a body is a measure of its energy content: if the energy changes by an amount L then the mass changes in the same sense by $L/9 \times 10^{20}$ if the energy is given in ergs and the mass in grams.

of space and time. Let us give an example of the uncanny effects that occur.

The most direct way in which particle physicists meet the effects of relativity is when measuring the lifetime of an unstable particle. Muons, copiously produced by cosmic rays and also at particle accelerators, fall apart after a rather short time (in about two millionth of a second). However, measuring the lifetime for a slow moving muon or a muon moving with high speed gives different results: the fast moving muon lives longer. It is a direct manifestation of the effects of relativity, and a fact of daily life at the particle accelerators. When the lifetime of a certain particle is reported one must specify its state of motion. In the tables used by particle physicists the lifetime is usually understood to be the time measured when the particle is at rest. There is a similar effect when measuring distances. The precise equation relating distance measurements was deduced by Lorentz even before Einstein introduced the theory of relativity; this is the reason why one speaks of a Lorentz transformation when relating quantities measured in reference systems moving with respect to each other.

Finally, from the discussion before, we know that a point on the mass-shell (corresponding to a particle of definite mass, momentum and energy) will transform under a Lorentz transformation to another point on the mass-shell. Precisely how x became x' as discussed above. The Lorentz transformation specifies precisely where the point x' will be, given x and the relative velocity of the systems. In this sense the mass-shell is an invariant: a particle of given mass remains on the same mass-shell when going to another reference system. For a particle of zero mass we have the light cone, and going to a differently moving system a point on the cone (for which energy equals the magnitude of the momentum) will become another point on the cone (where again energy equals the magnitude of the momentum). The values of energy and momentum however will of course be different.

We have thus a number of invariances in physics. The invariance of the laws of physics with respect to rotations and with respect to systems moving with a constant speed relative to each other is now generally called Lorentz invariance. Including invariance with respect to translations in space and time one speaks of Poincaré invariance. Both Lorentz and Poincaré made their contributions prior to Einstein; it is Einstein who invented relativistic kinematics and made us understand the whole situation in full clarity.

Invariance with respect to relative movement can be used with advantage to understand certain situations. If some physical process is forbidden (or allowed) in some system it is forbidden (or allowed) in systems that move relative to that original system. For example, if some decay process does not occur if a particle is at rest it will also never occur if it moves; we shall effectively use this seemingly trivial observation to clarify complex situations. Deducing things in the most convenient reference frame is often of great help in particle theory.

4.5 The Relation $E = mc^2$

The equation $E = mc^2$ is surrounded by mystique, and there is the folklore that this equation is somehow the starting point for making an atomic bomb. It might not do any harm to explain this equation in some detail, to demystify it.

In the simplest possible terms this equation means that energy has mass. Given that the weight of an object is proportional to its mass this means that energy has weight. Consider an old-fashioned watch, with a spring that must be wound regularly. When the spring is completely unwound, measure the weight of the watch. Then wind it, meaning that you put energy into the spring. The energy residing in the spring has some weight. Thus if you measure the weight of the watch after winding the spring it will be a little heavier. That weight difference is very small but non-zero.[c] You need really a massive amount of energy before the additional

---

[c]It is something like one hundred-millionth-millionth part of a gram.

weight becomes noticeable. A little bit of mass corresponds to a very large amount of energy. That is because the speed of light is so large. A radio signal goes seven times around the earth in one second.

Here another example. Take a car, weighing, say, 1000 kg. Bring it to a speed of 100 km/h. The weight of the corresponding energy is one half of the hundred-millionth part of a gram $(0.5 \times 10^{-8}$ gr$)$.[d] You can see that energy weighs very little; no wonder that nobody ever observed this effect before Einstein came up with his famous equation. It took a while (till about 1937) before it was demonstrated explicitly.

As yet another example consider a double sided cannon. This is a type of cannon that might be useful if you are surrounded, and that fires two cannonballs in opposite directions. Thus there is a long cannon barrel, and one inserts a cannonball at each end. You could imagine gunpowder between the two balls, but here we will suppose that there is a very strong spring that is pushed together. Once pushed as far as possible a rope is attached that keeps the two balls together. At the command "fire" some person cuts the rope and the two balls will fly off in opposite directions, with a velocity determined by the amount of energy stored in the spring.

The above figure shows this idea, the green line is the rope keeping the balls together. Measure very carefully the weight of the cannon barrel, the two cannonballs, the rope and the spring

---

[d]It can be computed by evaluating $\frac{1}{2}Mv^2/c^2$ where M is the weight of the car at rest in grams while v is the speed of the car, about $\frac{1}{36}$ km/s, and c the speed of light, 300,000 km/s.

before pushing in the balls. Then push in the two balls. That will cost you energy, and that energy will be stored in the spring. Next make again a measurement of the weight of the whole ensemble. The result will be that the total is now slightly heavier than with uncompressed spring.

A decaying neutron has much in common with our double sided cannon. To paraphrase Einstein, God throws his dice, and when a six comes up he cuts the rope. Thus when the neutron decays, two particles, an electron and a neutrino, shoot away (not necessarily in opposite directions), and a proton remains more or less at the place of the neutron. Here the energy is relatively large: the difference between the neutron mass and the sum of the proton and electron mass (the neutrino mass is very small or zero) is about 0.1% of the neutron mass. It translates into kinetic energy of the electron and the neutrino, the proton remains practically at rest. In particle physics Einstein's equation is very much evident in almost any reaction.

Now what about the atomic bomb? The function of the equation $E = mc^2$ is mainly that one can tell how much energy becomes available by simply weighing the various objects taking part in the process. A uranium nucleus becomes unstable when a neutron is fired into it, and it breaks up in a number of pieces (including several neutrons, which can give a chain reaction). The pieces are nuclei of lighter elements, for example iron. Since the mass of the uranium nucleus is well-known, and since the masses of the various secondary products are known as well, one can simply make up the balance (in terms of mass). The difference will be emitted in the form of kinetic energy of the decay products, and it is quite substantial. So that is what Einstein's equation does for you: you can use it to determine the energy coming free given the weight of all participants in the process. Perhaps it should be added that in the end the kinetic energy of the decay products will mainly translate into heat (which is in fact also a form of kinetic energy of the molecules). The real energy producing mechanism here resides in the way the protons and neutrons are bound together in the nuclei.

During his life Einstein used different methods to derive his equation. Originally he took the hypothetical case of an atom at rest emitting light in opposite directions, so that the momentum of the atom was zero both before and after the emission. Then he considered how this looks from a system moving with respect to this radiating atom. He knew precisely how the light rays looked in the moving system: for that one uses the light cone. Assuming conservation of energy he could state quite precisely what the energy difference was between the initial and final state of the atom. Thus he looked at it in two different systems: one in which the atom is at rest both before and after the emission, and one where the atom had momentum both before and after. He also knew precisely the difference in energy of the atom between the two cases, because that was equal to the difference of the energy of the light if the system is at rest and the energy of the light in the moving system. In other words, he got a piece of the curve, and from there his equation follows. To say it slightly differently: once he knew about the light cone, he could deduce what happened in other cases by considering what happens if light is emitted. Conservation of energy is the key to his derivation in all cases.

⟨5⟩

Detection

5.1 Introduction

In this Chapter the experimental methods of particle physics will be discussed in a cursory way, and not in depth; others are better qualified to do this. But it is necessary to have some idea how elementary particles are detected and observed. Also, it is very hard to resist the temptation to discuss the photoelectric effect; it is such a beautifully simple, easily described effect, and yet its consequences are immense: the particle structure of light.

The detection methods have changed grossly through the years. Before 1950 Geiger counters, photographic emulsions and Wilson's cloud chamber were the major detection instruments used; after that the bubble chamber took over much of the task. In the early sixties the spark chamber made its entry, and evolved to what is called the proportional wire chamber. In addition, today, semiconductor (the same material as used in chips) strips are used to detect particles.

The principle of many detectors is the detection of a track left by the passage of a charged particle. Hence only charged particles can be observed in those detectors. When a charged particle passes through matter, it knocks out electrons from the atoms, thereby disturbing the structure of the material, and also creating loose electrons. Thus a charged particle passing through matter leaves a trail of disturbed matter, of ions, and loose electrons that can be collected. An ion is an atom or molecule with one or more missing (or extra) electrons. The ions along the path usually lack

Charles T. R. Wilson (1869–1959). He invented the cloud chamber. As early as 1895 he discovered that water vapor would condense around charged particles. Measuring the charge of such droplets was the method of choice whereby particle charge was measured.

Wilson kept on working and by 1911 he had developed his cloud chamber. This chamber made particle tracks visible by water vapour condensing around the ions, and he did photograph them. As vapour is not very dense, this instrument was not suitable to observe particle reactions where the material of the detector functions as target. There is simply not enough target material. Photographic emulsions and bubble chambers (containing liquid) are more suitable for that. In 1927 Wilson shared the Nobel prize with Compton (1892–1962).

Cloud chambers and emulsions became less important to experimental physics after the invention of the bubble chamber by Glaser. However, they have been quite instrumental in the development of particle physics. And let us not forget that Glaser acknowledged the cloud chamber as his starting point.

Cloud chambers are not difficult to construct. On the web you can find drawings and manuals for making such an instrument. It will allow you to see cosmic rays and discover many interesting things. You can also see tracks from radioactive sources, in particular if they emit alpha-particles. An alpha-particle is a combination of two neutrons and two protons, in fact precisely a helium nucleus.

one or more electrons, and are thus positively charged. The figure below gives an idea of how a track looks like: the ions (blue dots) will not move very much, the knocked off electrons drift away (little red lines with arrow).

Charged particles passing through certain organic materials may produce visible photons. This was originally discovered[a] for naphthalene, the stuff that mothballs are made of. What happens is that in these complex molecules electrons may be kicked into higher orbit. Next these electrons fall back to their original orbit and the energy released then is emitted in the form of light, photons. This light can be seen as a very short, blue tinted flash. The active material is usually dissolved in a liquid. Scintillation counters are based on this effect: a charged particle passing through such material produces a small flash of light that can be observed using photomultipliers (see below). Obviously this works only if the material used is transparent to the light produced, so that this light can be detected outside the material. Scintillation counters are often used as trigger or anti-trigger. In the first instance, a track observing apparatus would be activated only after a scintillation counter had shown the passage of a particle, in the second case one may be interested in cases where at certain places no particle moves, and then a scintillation counter causes the detection apparatus not to be activated when a particle passes through that particular counter.

Photons passing through matter also cause observable effects. If a photon is of sufficiently high energy it causes pair-production (production of an electron and a positron) when passing through

---

[a]Remarkably, this discovery was made at the end of the war by Kallman and his student Broser. Kallmann, Jewish, miraculously survived the war in Germany while continuing his experimental work.

Pavel Cherenkov (1904–1990), **Il'ia Frank** (1908–1990) and **Igor Tamm** (1895–1971), physics Nobel prize 1958. Cherenkov (and also Vavilov) discovered what is now called Cherenkov radiation, while Frank and Tamm developed the corresponding theory. The discovery is a wonderful example of an experimental discovery. In those days (1934) fluorescence was a commonly known phenomenon. It amounts to absorption of some kind of radiation (actually also sound waves can do it) by some materials, followed by subsequent emission of light. That light may actually be emitted substantially later in time. Fluorescence was studied by many, especially Becquerel (father and son) were great experts on that, and this played a role in the discovery by Becquerel of radioactivity. In the case at hand Cherenkov studied the effects of gamma rays (these are photons) emitted by a radium source and passing through some solvents. He saw a faint bluish light, which upon further study was caused by fast electrons produced by the gamma rays interacting with the molecules of the fluid. Initially, as Cherenkov stated, this seemed of no special interest, since it appeared to be just fluorescence, widely studied before among others by the Curies. In fact, the bluish light had been seen before.

Only after very detailed investigations Cherenkov established the true nature of this light: it was due to the passage of very fast electrons through matter. The theoretical investigations of Frank and Tamm elucidated the mechanism. It was caused by the fact that the speed of the electrons was larger than the speed of light in the medium. It is amusing to note that much earlier, in 1904, Sommerfeld (a brilliant physicist and teacher, with notably Heisenberg and Pauli as students) had already considered the problem in another context. At that time, before Einstein, the velocity of electrons was theoretically not limited by the speed of light, and Sommerfeld discussed the 'sonic boom' produced by such fast moving electrons.

Interestingly, very likely Cherenkov did not build any Cherenkov counter himself. These devices are now an important part of almost every particle physics experiment.

the electric field (Coulomb field) near a nucleus. The pair subsequently produces tracks that can be observed. High-energy electrons passing near a nucleus often emit photons. This is called bremsstrahlung, which is the German way of saying brake radiation. The electron brakes in the electric field of the nucleus. If the electron is of sufficiently high energy, the photons produced may be sufficiently energetic to make an electron-positron pair in the electric field of some other nucleus. These particles in turn produce again bremsstrahlung, and the result is an avalanche, commonly called a shower. Showers, caused by highly energetic photons, electrons or positrons are very characteristic of these particles.

The electrons and positrons produce tracks and can be seen, while the photons themselves do not make a track. The number of electrons and positrons is a measure of the initial energy of the incident particle. In the drawing the dashed lines are the photons. Charged particles much heavier than the electron (in practice this means all other particles) produce much less bremsstrahlung and hence do not give rise to showers.

Yet another way to observe particles is through Cherenkov radiation. A charged particle passing through water, for example, emits visible photons, much like an airplane produces sound. Now light in water, winding its tortuous way through the liquid, propagates with a speed less than the speed of light in vacuum. A fast high-energy particle moving through water may thus move with a speed higher than the speed of light in that medium. The result is the optic equivalent of a sonic boom. The particle leaves behind an expanding cone of light. This radiation is called Cherenkov radiation, after the Russian physicist who discovered it. It is usually of a bluish color, and can be seen quite clearly when looking into a heavy water nuclear reactor. The opening angle of the cone of radiated light depends on how much the speed of the particle exceeds the speed of light in that medium, and may thus be used to determine the velocity of the particle precisely. One uses photomultipliers to observe the Cherenkov radiation.

Neutral particles (photons, neutrons) themselves do not produce an ionized trail, therefore they can be observed only indirectly. Photons are observed through the electron-positron pairs they produce in matter. Neutral particles that have strong interactions (such as neutrons) usually collide quite quickly with some nucleus (with often the nucleus breaking up) and that will generally give rise to several charged particles, or even nuclear fragments. Finally, a neutral (or charged) particle may be unstable and decay, and if its decay products are electrically charged they can be observed.

Let us summarize the various methods.

Fast moving charged particles ionize matter and this disturbance can be observed in various ways. Geiger counters, Wilson cloud chambers, bubble chambers and spark chambers operate that way. Proportional wire chambers observe the electron avalanches created by electrons kicked out of the atoms.

Charged particles may excite certain molecules, and when these de-excite they emit light. Scintillation detectors are based on this principle.

Highly energetic charged particles passing through a medium may produce light much like a sonic boom of an airplane flying with a speed exceeding that of sound. Cherenkov detectors are based on this principle.

Low energy photons hitting matter cause electrons to be kicked out of that material. This is like a stone splashing into water, causing water droplets to be kicked up. This is called the photoelectric effect, and photomultipliers are exploiting this mechanism. Only relatively low energy photons (visible and ultraviolet light) can be observed this way.

Neutral particles are detected indirectly; photons because they produce electron-positron pairs when passing near a nucleus. Also, neutral particles may collide with a nucleus, thereby breaking it up producing charged fragments. Some neutral particles are unstable and may decay into charged particles that can then be observed.

In addition it may be mentioned that these days semiconductor strips are used to detect the passage of a charged particle.

Human ingenuity produces constantly new ways to observe particles passing through matter. Methods come and go with time. Wilson cloud chambers, photographic emulsions and bubble chambers have all but disappeared; scintillation counters, spark chambers and proportional wire chambers dominate today's detection apparatus. Bubble chambers and spark chambers produce photographic pictures showing the tracks and that is, if nothing else, direct and suggestive. Proportional chambers pass their output directly to computers, and are in a sense less direct. Nowadays experiments are almost entirely run by computers. The pessimist might think that the time is near when computers will also publish the results, or rather pass them on to other computers, but such a view totally underestimates human ingenuity and the very human drive and thirst for knowledge. It might be that high-energy particle physics has become "big science", but in a sense that may be compared with space research. What does it take to go to the moon, or the other planets? That is hardly an individual enterprise, yet it must be done. We want to know!

In the rest of this Chapter we discuss some of the detectors in more detail: photomultipliers, bubble chambers, spark chambers and proportional wire chambers.

5.2 Photoelectric Effect

Photon detectors are a class apart. Photo tubes were invented in the very beginning of the twentieth century, and their behaviour was not well understood until Einstein, in the same year (1905) that he published his theory of relativity, suggested that light is quantized and the "photons" for light of a given frequency (color) have a well defined energy. This was subtly different from Planck's work, in 1900, who suggested that light is emitted by matter in certain well-defined packets of energy only. In other words, Planck suggested that the sender of light emits in quantum packets, Einstein on the other hand proposed that light itself can only exist in certain energy packets. It is Planck who introduced the relationship $E = h\nu$: the energy of a packet, a photon, is proportional to the frequency of light. Given the frequency of light (the quantity ν) and knowing the constant h one can compute the energy of the basic energy packet. The quantity h is called Planck's constant. Its value is $h = 6.626 \times 10^{-34}$ joule sec, deduced by Planck to fit the observed spectra of light emitting black bodies. Planck did not dream of interpreting this as a property of light, that was Einstein's contribution.

Let us repeat here an important fact. The energy-frequency relation of Planck is such that the energy packets, the photons, have higher energy as their frequency is higher. Blue light contains photons of higher energy than those found in red light, which has lower frequency. The energy of photons of visible light is very low on the scale of the things discussed in this book. Red light, for example, has photons whose energy is about 1.5 eV, for blue light the photon energy is 3 eV. Ultraviolet light contains photons of more than 3 eV, and X-rays have photons of even higher energy: hard X-ray radiation has photons with an energy in the 10 keV (1 keV = 1000 eV) range.

The introduction of Planck's constant is the very beginning of quantum mechanics. Indeed, this constant is now the universal basis of all quantization. Very fittingly, it was discovered in the year 1900. Let us quote from Pais in his book *Inward Bound*: "Were I asked to designate just one single discovery in twentieth century physics as revolutionary I would unhesitatingly nominate Planck's of December 1900." This discovery was really something. It was not "in the air", and no one else even vaguely suspected anything like it.

The photoelectric effect is this. When light hits a surface (the cathode) it may kick an electron out of that surface. By applying an electric field this electron may be drawn to another electrode, the anode, and thus gives rise to a very, very small current.

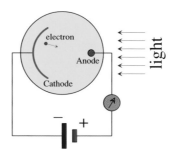

The puzzle was that the effect would occur only for light of a color on one side of some specific color in the spectrum as seen in a rainbow. In the rainbow one distinguishes the colors red, orange, yellow, green, blue, violet (it is of course a continuum), and the effect would for example occur for blue and violet light, but not for red, orange or green light. Using green light, no matter how intense, no electron would get kicked out, while the smallest intensity of blue light would show the effect. Einstein solved the puzzle by suggesting that light was not only produced in certain energy packets, but that it actually always came in the form of energy packets. He thereby introduced the idea of a photon. That is really a difference: emitting light in packets one may still build up the energy to any amount for a given ray of light. If light is always in the form of photons then there may still be any amount of energy in a beam of light, but it is always in the form of these little packets, the photons. That is like the difference between one big man and a hundred small men. Now photons are energy packets whose energy depends on the color (frequency). To kick an electron from material one needs a certain amount of energy, and then for example a blue photon could and a green photon could not kick an electron out of the surface because the green photon has less energy. That threshold effect is independent of the amount of light (the number of photons) projected onto that surface. Of course, once the photons are of the right color the current produced would be proportional to the intensity of the light, i.e. the number of photons. But to get anything to begin with one needed blue or violet light, and green light would just give nothing at all.

Let us quote some numbers here. To kick an electron out of the material one must overcome some threshold. The energy needed to cross the threshold, to get out of the material, depends on the material used. Now suppose that to overcome this threshold for some specific material an energy of 2 eV or more is needed. Then evidently the photons in red light, having an energy of about 1.5 eV, are simply not sufficiently energetic to kick an electron out of the material. The photons of blue light (3 eV) however can, and there is an effect. After coming out of the material the electrons still have some 1 eV energy left in the form of kinetic energy. It is an important test of Einstein's idea that this surplus energy goes up linearly with the frequency of the light used. In 1915 Millikan, through diligent research spanning several years, verified this fact, and used it to deduce Planck's constant with a precision of 0.5%. Remarkably, even then Millikan refused to accept the photon idea.

Great progress, technically, was achieved with the introduction of photomultipliers. Contrary to the name, what gets multiplied is not the photon in the light, but the emitted electron. That electron is, using electric fields, accelerated and when it hits the anode it will make a splash so that several electrons are kicked out. These are then accelerated again and directed to a second anode where then they again make a splash. Etcetera. In this manner a single electron produces an avalanche. Using this technique phototubes are now so sensitive that they can detect a single photon (with finite efficiency).

Donald Glaser (1926) invented the bubble chamber. He started with a 3 cm³ glass vessel filled with diethyl ether at the University of Michigan. It evolved quickly to large dimensions and the other picture above shows the 3.7 m Big European Bubble Chamber (BEBC) in retirement at CERN. It was filled with liquid hydrogen (thus kept at a temperature of −253°C). For most reactions seen the target was thus simply a hydrogen nucleus, i.e. a single proton. Furthermore, using hydrogen gives clean and sharp tracks, as can be seen in the picture on the next page. A disadvantage is that photons will in general not convert to a shower inside a hydrogen filled chamber. An electron will not convert to a shower either.

The bubble chamber dominated experimental particle physics for quite some time. Heavy liquid bubble chambers were filled with freon (the liquid used in refrigerators); they were used if much target mass was required. Also, photons convert readily into a shower in such chambers. Literally millions of photographs were taken in hydrogen, propane and heavy liquid-filled bubble chambers.

Rumour has it that Glaser got the idea when staring at the bubbles in a glass of beer in a pub in Ann Arbor called the Brown Jug. I asked him, but he denied it, although at one point he tried beer as a possible liquid. His basic starting point was the Wilson cloud chamber. Of course, the bubble chamber, having much more mass (that can function as a target), was more suitable for particle physics. In 1960 Glaser was awarded the physics Nobel prize.

Don is quite a ladies' man. Combine that with a Nobel prize and you have an explosive mixture.

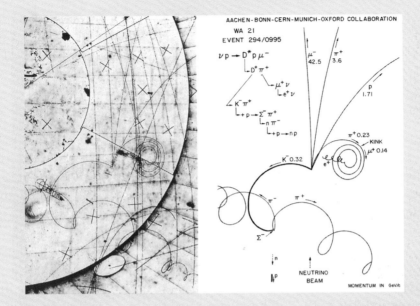

The photograph above shows an event in the Big European Bubble Chamber (BEBC). The event was caused by an incoming neutrino colliding with a proton. The basic reaction is with a down quark in the proton (that contains two u and one d quark):

$$\text{neutrino} + d \rightarrow \text{neg. muon} + c$$

At this point we have (apart from the muon) two u quarks (remainder of the proton) and a c quark. Out of the glue mass in the proton a down–antidown pair is created. The antidown quark combines with the charm quark to make a bound state of charge +1 and called the D^*. That leaves one down quark and two up quarks, which is again a proton, and we have the reaction:

$$\text{neutrino} + \text{proton} \rightarrow \text{neg. muon} + D^* + \text{proton}$$

The c quark is not stable, and the D^* decays in a complicated way, in a very short time, to two positive pions and a negative K-meson. This then is what is seen at the pimary vertex:

$$\text{neutrino} + \text{proton} \rightarrow \text{neg. muon} + \text{proton} + \pi^+ + \pi^+ + K^-$$

The K^- causes another reaction further down.

Only an experienced person can figure this out, using theory. For example the D^* lives too short a time to be seen explicitly. Conservation of energy-momentum must hold. The momentum of each particle follows from the curvature of its track and is given in GeV. The energy of the initial neutrino is not known a priori.

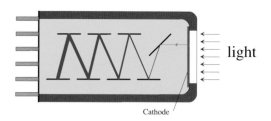

Photomultipliers are nowadays used mainly in two ways: as detectors of the light flashes of scintillation counters and as detectors of Cherenkov radiation. Photomultipliers cannot be used to detect photons with an energy beyond the far ultraviolet.

5.3 Bubble Chambers

Bubble chambers have dominated high-energy experimentation in the period 1953–1973. They were invented by Glaser at the University of Michigan in 1952. They can be seen as a natural evolvement of cloud chambers, and to a large extent they are based on the same principles. The mechanism is this. Consider a liquid such as water. The boiling point of water depends on the pressure of the surrounding air, and for example up in the mountains water boils at a much lower temperature than at sea-level (where it boils at 100°C). Now fill a chamber with water, with a piston that can be moved, so that the pressure of the air above the water surface can be varied. First push the piston down, and maintain the temperature at such a level that just no boiling occurs. Expose the chamber to particles. These particles leave a track of ions (and loose electrons). Now move the piston up: the pressure decreases, and the liquid will start to boil. It will in fact start to boil first along the perturbation: many little bubbles appear along the path. Take a picture, and move the piston down again before overall boiling occurs. The figure shows a schematic drawing of a bubble chamber (including a magnetic field, see below).

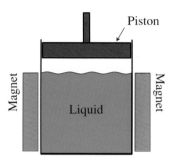

In many experiments the bubble chamber liquid is at the same time the target for the beam of particles entering the chamber. The beam particles hit the nuclei in the liquid, and as a consequence of the collisions other particles can appear and produce tracks that will show in the picture. The target liquid may be chosen to fit the particular processes that one wants to observe; there exist for example liquid hydrogen bubble chambers that operate near the boiling point of hydrogen (−253°C). Others contain heavy liquids for the case that much target mass is required.

Most bubble chambers have a strong homogeneous magnetic field over the volume of the chamber. Charged particles moving through such a field will curve one way or the other depending on the sign of the charge. The magnitude of the curvature depends on the momentum of the particle; the path of a fast particle curves much less than that of a slow particle.

Millions upon millions of bubble chamber pictures have been taken. They required much effort to analyze: the pictures had to be scanned and the tracks measured. Most of that was done semi-manually, because it is very hard to fully automate the process of track recognition. There is always a lot of stuff on such a picture, uninteresting events, beam particles going through without doing anything, spontaneous bubbles not associated with particle tracks etc. Many particles were discovered in these pictures, and in the early seventies a certain type of event (a collision is usually called an event) was crucial in the verification of gauge theories. These events were obtained exposing a very large heavy liquid bubble chamber to a neutrino beam. The events were neutral current events, to be discussed later.

The next figure is a reasonably faithfully reproduced event observed in a heavy liquid bubble chamber. As with all events, nothing is absolutely sure, but here follows its latest interpretation. The event was seen in a neutrino run, and the incident particle was a neutrino, entering from the right. That neutrino, not producing any track as it is neutral, collided with a neutron in some nucleus, which is the starting point of the event. The fat short track moving upwards and to the right is the recoiling nucleus. The track is fat because the charge of that nucleus is high as it will contain many protons in this heavy liquid (freon). This strongly ionizing track is marked with an 'N'.

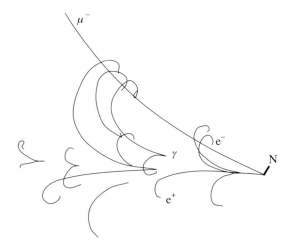

The long, upward and slightly curved track is a muon (marked with μ^-); for all practical purposes it has only electromagnetic interactions and interacts very little with the liquid of the bubble chamber; being much heavier than the electron it is much less likely to produce bremsstrahlung. This is typical for a muon: a long track showing no interactions. That is also how one detects muons in general: after a reasonable amount of matter has been traversed only muons (and of course neutrinos) remain. The curvature, clockwise, is that of a negatively charged particle. The muon has a comparatively long lifetime and will usually leave the bubble chamber before decaying. Other particles, such as protons, neutrons or pions (not present in this picture) have strong interactions, i.e. will interact with a substantial probability with some proton or neutron of some nucleus, also often within the bubble chamber. That type of interactions is fairly typical: generally various particles produced in the collisions are emitted more or less uniformly distributed over all directions.

The remaining visible tracks are probably all electrons or positrons. One of the electron tracks, curving clockwise, has been marked e^-, and an anti-clockwise curving track, most likely a positron, has been marked e^+. All these particles moving through the liquid lose energy while ionizing molecules along their paths. For this reason the particles will slow down and the curvature of the tracks increase; eventually the particles come to a halt.

A high-energy photon, usually called a gamma ray, may convert with some probability to an electron-positron pair in the electric field of a nucleus. One such pair has been marked with γ in the figure. The average distance that a photon covers before converting depends on its energy and the material traversed. There is a probability distribution, hence the distance varies from case to case. In a heavy liquid bubble chamber the probability of such an interaction is quite high, and one will detect most of the photons; on the other hand a photon has to cross some distance before interacting, so one will observe a gap as the photon itself is invisible in a bubble chamber. The electron-positron pair has the form of a V and points in the direction from which the photon came. The energy of the electron in a pair need not be the same as the energy of the positron, thus their tracks may be different in curvature. It can even happen that one of the tracks is so short that it is next to invisible, which is one of those misleading things that may occur. The single track, bottom, slightly left of the middle, may be of that type, but there are other possibilities. This event, one of the very first observed in the ill-fated 1963 CERN neutrino

Shuji Fukui (1923) and **Georges Charpak** (1924) invented the spark chamber and the multi-wire proportional counter respectively. Both inventions have been and still are of tremendous importance in particle physics. In particular, Fukui's invention, the spark chamber, was absolutely crucial to the neutrino experiments in the early sixties. Without them the Brookhaven neutrino experiment would have been impossible. It is one more example of how technological advance leads to an advance in physics, in this case quite immediately to the discovery of two neutrino species (the electron-neutrino and the muon-neutrino). The picture shows Fukui in the counting room at the 1963 CERN neutrino experiment.

The proportional wire chambers are of crucial importance in today's experiments. They allow a much more digitized setup, capable of handling much higher event rates. They are crucial to experiments at the LHC (Large Hadron Collider), starting operation in 2007 or so.

Fukui is a rather shy person, who for one reason or another has not been recognized that well. This in contrast to Charpak, who received the 1992 physics Nobel prize. Charpak is an ebullient character who has become a TV personality in France. He is now not very sure if that is good or bad. In any case, he told me that he is taken to be an expert on anything from condoms to nuclear energy. This latter point has got him into some trouble, so once more, if you get a Nobel prize do not pretend to know everything even if the media seem to think so.

experiment, has never been understood with any certainty. At the time this event, essentially containing only electrons, positrons and a muon, generated quite some excitement, but no other event anywhere like it was observed afterwards. It was generally known as the Agnes event, after the French physicist Agnes Lecourtois, who made a very careful study of it.

5.4 Spark Chambers

Spark chambers, invented by the Japanese physicist Shuji Fukui, consist of parallel metal plates with a certain gas in between those plates. When a particle passes through that gas it leaves a trail of ions and loose electrons. Applying now a very high well chosen voltage to the plates a spark will develop along the string of ions along the path. In this way a track becomes visible as a series of sparks, and a picture can be taken. Again, the material of the plates may function as target for particle beams, and one studies then the reactions of the beam particles hitting the nuclei in the plates. In the figure the incident particle is a neutral particle, and thus produces no track until it hits a nucleus in a plate (here the third), producing two secondary charged particles.

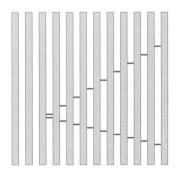

There are several advantages to spark chambers over bubble chambers. First of all, they can be triggered (applying a pulsed voltage) with great precision. Scintillation counters may be used to establish whether a given event is of a type to be investigated;

Frank Linde (1958) and **Jos Engelen** (1950), experimental particle physicists of the University of Amsterdam and the NIKHEF (Dutch particle physics Institute in Amsterdam) with a sampling calorimeter and crystal calorimeter respectively. The sampling calorimeter (left) consists of a number of thin heavy plates (iron, lead or uranium) separated by plates of scintillation material. The heavy material slows down the incoming particle(s) and when they pass through the scintillator material light is produced. This light is guided through plastic plates and collected at the spot pointed at by Linde. A phototube registers the intensity of the light, which depends on the number of scintillator plates activated.

The crystal calorimeter (right) is used for electron and photon energy measurements. This very expensive instrument consists of crystal material (the reddish stuff visible on the left half) in which the electrons and photons produce showers that again produce scintillation light registered by means of phototubes. In these pictures Engelen (actually now the director of the NIKHEF) looks at the crystal calorimeter mainly used by Linde at the DESY laboratory in Hamburg, Germany, while Linde stands besides the calorimeter used by Engelen.

Calorimeters have become indispensable at the big accelerators. In a way they have taken the place of magnetic fields that would bend particle trajectories depending on their energy. At very high energies the curvature is very, very small, so that the magnetic field method becomes useless. Furthermore, the calorimeters can measure the total energy of whole bunches of particles (particle jets).

then only for these events the chamber is triggered. So one obtains pictures that contain almost exclusively the type of events of interest in a given experiment. Secondly, by using many relatively thick metal plates, the target can be made quite massive, which is necessary for the study of such elusive particles as the neutrino.

Another important instrument is the calorimeter. The sampling calorimeter consists of heavy plates separated by detectors. These detectors could be spark chambers or scintillators The materials used in the heavy plates must be sufficiently dense so that if certain very high-energy particles (not a neutrino, but for example a proton or a neutron) pass through the plates a collision with a nucleus will be likely. Thereby the particle slows down. The products of the collision, still quite energetic, will go on and produce further reactions. All this then gives rise to many sparks or scintillation flashes, and the amount of light emitted becomes a measure for the energy of the incident particle. Careful design results in rather accurate energy measurements.

5.5 Proportional Wire Chambers

F. Krienen, an engineer at CERN, developed a method to digitize spark chamber events. He also invented digitized wire chambers with magnetic core readout. This development was carried further to the proportional wire chamber by G. Charpak, also at CERN. A proportional wire chamber exploits the loose electrons along a track. Instead of plates as in the spark chamber one has many wires at small distances strung in a frame. There are voltage differences between the wires. When a particle passes through the frame the electrons produced along the track develop small avalanches which will drift to the wires. Upon arrival there they give rise to very small currents that are very precisely measured and timed. The total information obtained this way can be used to determine the track with a precision of a fraction of a millimeter in the direction orthogonal to the wires, with a timing precision of 10 nano-seconds (a micro-second is one-millionth of a second, a nano-second is one-thousandth of a micro-second). One uses many frames oriented in different directions to obtain a precise measurement of the track of a particle. There is an enormous amount of electronics associated with this (every wire needs an amplifier etc.), and furthermore the events observed are usually directly processed through computers. No pictures are taken.

6

Accelerators and Storage Rings

6.1 Energy Bubbles

A new concept must be introduced here, one that will make it easier to understand particle production and decay.

The concept is that of an energy bubble. Think of it as a very, very small area of concentrated energy, of nondescript form but with nonetheless some specific properties. These are the values of the quantum numbers, the energy and the momentum of the bubble. For example, a bubble may have charge two, have a total energy of 200 GeV (1 GeV = 1000 MeV) and zero momentum.

An energy bubble will decay with a finite probability into any combination of particles that is allowed given the conservation of energy and momentum and the conservation of quantum numbers.[a] The probability with which the bubble decays into any given configuration differs from configuration to configuration. For a given bubble and a given configuration that probability though is always the same.

It is important to realize that the energy in the bubble available for particle production does not include the kinetic energy of the bubble itself if that bubble is moving. Such a moving bubble is like a bubble at rest observed from a moving laboratory. Therefore a moving bubble cannot ever decay into some configuration of particles if that decay is not possible for the same bubble at rest.

---

[a]In actual fact there are yet further conservation laws, such as the conservation of angular momentum, but a discussion of those would be counter-productive here.

The kinetic energy of the bubble itself plays no role in the decay process.

Elementary particles themselves are energy bubbles with well defined quantum numbers. In addition they are on the mass-shell, as explained before: there is the stringent relation between the momentum and the energy of the particle. So that is how we must see an elementary particle: an energy bubble with a well-defined mass. Other than that there is no difference. In particular it will decay if that is permitted by the conservation laws. The decay may be slow or fast depending on the possible decay modes and the probability of decay for each of these decay modes. Some particles take so long to decay that they may traverse significant distances, and then they can be observed directly, for example in a bubble chamber. An example is the muon, which needs about one millionth of a second to decay. That is enough[b] to have it traverse quite large distances, in fact enough to make it go around in storage rings. Other particles decay so fast that they show no appreciable track length; in such a case the identity of the particle must be established by investigating the specific decay products of such a particle and see if the total energy and momentum of the decay products add up to values that are on a mass shell. Thus look at many cases, and see if some specific combination of particles is always on the same mass-shell. Given energy and momentum the associated mass is easily found: take the graph for the connection between energy and momentum given before and plot the point corresponding to the given momentum and energy. Do this many times and see if this produces a piece of a mass-shell curve. If so, one has found a new particle, decaying into the specific combination studied. Traditionally such particles were often called resonances, but here they will not be considered to be anything different from longer-lived particles.

---

[b]According to the theory of relativity fast moving particles will live longer, and for energetic muons this effect greatly increases the distance traversed before decay.

Obviously all but the relatively low mass particles are unstable and will decay. Apart from the zero mass particles the stable particles are the electron, the proton and their antiparticles. The proton is of course stable thanks to the non-zero baryon number. That is like the electron, not decaying to photons, or neutrinos: charge must be conserved, and the only particles with masses below the electron mass are the zero mass particles, all electrically neutral. The baryon number is another quantum number, and the proton has baryon number 1. There are no stable particles with non-zero baryon number lighter than the proton, and therefore the proton is stable. The neutron can and does decay into a proton plus electron and antineutrino. Note that charge conservation holds here: the initial state, the neutron, has charge zero, and the final state, with a proton (charge 1) and an electron (charge –1) has zero charge as well.

Neutral particles produce no tracks, and therefore they can be observed only indirectly. Those neutral particles that have strong interactions usually collide quite quickly with some nucleus (often with that nucleus breaking up) and generally that will give rise to several charged particles. In other cases, the neutral particle may decay, and if the decay products are electrically charged those can be observed. The photon is a special case: when passing through matter (such as a gas or a liquid) it causes production of single electrons (kicked out of an atom) or electron-positron pairs that can be observed.

Production of (new) particles is a matter of producing an energy bubble of as high an energy as possible. Such a bubble comes into existence as a result of a collision. All you have to do, after producing such a bubble, is to wait and see what comes out. With a certain probability anything compatible with the conservation laws will be produced, although sometimes that probability may be so small that the process becomes essentially unobservable. In any case, here is the basic recipe: accelerate some particle to the highest possible energy and then have it collide with another particle. See figure below. This is what is done at the various high-energy laboratories

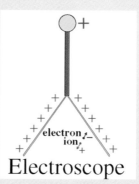

Victor Hess (1883–1964, Nobel prize 1936) and **Theodor Wulf** (1868–1946). These physicists discovered cosmic rays. In 1910 Theodor Wulf, a Jesuit College teacher in Valkenburg, the Netherlands, made a sensitive electroscope (Wulf's electroscope). It was known that such an instrument, after being charged, slowly lost its charge and it was believed that this was due to radiation from the earth. It was known at the time that radio-activity would discharge such an instrument. Wulf asked the French physicist Langevin for help to do the experiment at the top of the Eiffel tower. The result, carefully analyzed, was unexpected: the electroscope discharged much faster than anticipated given the absorption of radiation by the air!

An electroscope is a very simple device of which the main part consists of two conducting leaves. When charging this setup the leaves will repel another, and they will spread out, as in the picture. If a charged particle passes by, knocking off electrons from atoms, the resulting ions or electrons drift to the leaves, thereby discharging them, and they fall back.

Hess decided to investigate the issue in a systematic manner. He started off with some experiment in a meadow in Vienna. In order to get higher up he became a balloonist, taking Wulf's electroscope to heights of up to 5 km. After some 8 flights (sometimes unmanned), a few of them at night and one during a solar eclipse (to eliminate the sun as a source) he established that at high altitudes the effect was stronger than near the ground, concluding that the effect was due to radiation from outer space. Millikan entered the field later on, and having a better sense of public relations coined the name cosmic rays (replacing the name ultra-radiation). At first, on the basis of his own experiments, Millikan doubted Hess's results, but later on he turned around, and in fact became more prominent in the public eye than Hess. The Swedes however recognized the facts and awarded half of the 1936 Nobel prize to Hess for the discovery of cosmic rays (the other half to Anderson). Perhaps they should have included Wulf.

around the world. Historically, the development of accelerators has been nothing short of stunning, and progress in elementary particle physics has advanced thanks to the construction of these accelerators.

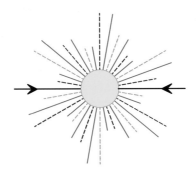

6.2 Accelerators

The biggest accelerator of them all comes for free: the Universe. In 1911 cosmic rays were discovered by Hess in Austria (Nobel prize 1936) following the experiments of Wulf on the Eiffel tower. It is now understood that particles with energies exceeding by far the energies achieved at man-made accelerators come to us from the Universe. The highest energies measured in cosmic rays reach 10^{21} eV, which is about a thousand million times (10^9) as much as the highest man-made energy (2 TeV, remember 1 TeV = 1000 GeV; 1 GeV = 1000 MeV; the electron mass is 0.5 MeV, the proton mass about 1 GeV). These cosmic ray particles are somehow produced and accelerated in the Universe and then traverse enormous distances to finally hit the earth. No one really knows where they come from or how they obtain their energy. Most of them collide at very high altitude in the atmosphere producing a cascade of secondary, tertiary, etc. particles which is then observed at the earth's surface as a shower that may cover quite a large area. Cosmic rays have had their place in the development of particle physics, and even now there are big experiments running or under construction making use of cosmic rays. The positron was first seen in cosmic rays, in 1932, by Anderson at Caltech. The muon and the pion as well as the K-meson were also discovered in cosmic rays, around 1950.[c] The trouble with cosmic rays is that the experiments are not under complete control of the experimenters. No one knows the identity and the energy of the initial particle causing an event at the earth's surface. It may be the original particle coming out of the Universe, or it may be one of the secondaries. This makes it very difficult to do systematic research. That is a general problem with experiments not under complete control. Observation of solar neutrinos or neutrinos from cosmic rays fall into this class, and to this day these experiments have a hard time clarifying the issue at stake (the issue here is whether neutrinos mix and/or have a mass).

Again, anyone interested in the fascinating subject of particle accelerators should read other books. Here only rough outlines are presented.

The tubes that display pictures in your TV or computer contain in fact an accelerator. In these tubes electrons are accelerated and deflected, to hit the screen thereby emitting light.

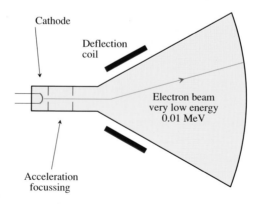

---

[c]The pion and the kaon (or π-meson and K-meson) are not elementary particles, but low mass bound states of quarks and antiquarks, which of course no one realized when they were discovered. They will be discussed in more detail in Chapter 8.

In the tube, on the left in the picture, a piece of material called the cathode, is heated. As a consequence electrons jump out of the material, and if nothing was done they would fall back. However, applying an electric field of several thousand Volts they are pulled away and accelerated. Then they are deflected, usually by means of magnetic fields generated by coils, deflection coils. They make the beam of electrons move about the screen. There they create the picture that you can watch if you have nothing better to do.

Thus inside your TV tube there is an accelerator, although of rather low energy as accelerators go. The energy of the electrons is expressed in electron-Volts. An electron has an energy of one electron-Volt (abbreviated to eV) after it has been accelerated by passing through an electric field with a potential difference of 1 Volt. In a television tube the potential difference may be something like 10 000 V, thus the electrons in the beam, when they hit the screen, have an energy of 10 000 eV, i.e. 10 keV. In particle physics one uses the unit MeV, one MeV = one million eV, and it follows that the electrons inside the TV tube have typically an energy of 0.01 MeV.

The starting point of any proton accelerator is the ion source. That may be a chamber filled with hydrogen. The simplest atom is the hydrogen atom. It contains just one proton as a nucleus with one electron orbiting around it.

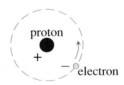

Next strong electric discharges are produced in this chamber. This causes the atoms to be ionized, that is the electrons are kicked out of the atom. The remainder of the atom is called an ion, and in the case of hydrogen that is simply the proton. The result is a collection of free electrons and protons along the path

John Cockcroft (1879–1967) and **Ernest Walton** (1903–1995), Nobel prize 1951. They constructed the first accelerator useful for doing nuclear physics experiments; their first observed reaction was the splitting of a lithium nucleus by means of a proton. The lithium nucleus contains 3 protons and four neutrons, and together with the incident proton one obtains a pair of alpha particles, each alpha particle containing 2 protons and 2 neutrons. It is amusing to read how they established this reaction. They had to demonstrate that a reaction gave rise to two alpha particles being emitted simultaneously. They did that using two observers each watching a screen that would light up if hit by an alpha particle. If they saw such a light flash they would tap a key. Two coincident keystrokes would indicate a lithium nucleus disintegration.

The machine they used for this experiment accelerated protons to about 700 000 Volts (700 keV). By today's standard this is of course a rather low energy, but it is already a quite useful energy for the purposes of the study of atomic nuclei. Optimistically, Walton thought as late as 1951 (see his Nobel lecture) that this method could be used to acceleration of protons to about 10 GeV (i.e. 10 000 MeV or 10 000 000 keV), but that was certainly a vain hope. Circular accelerators such as cyclotrons etc. are better suited for proton acceleration. Interestingly, a Cockcroft-Walton type voltage generator (cascade generator) can be found in most TV sets, to generate the required voltage of 10–30 keV. Also, Cockcroft-Walton machines are used to this day as an initial accelerator just following the ionization chamber of large proton accelerators.

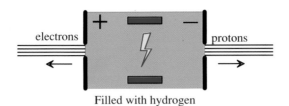

Filled with hydrogen

of the discharges. If nothing was done the electrons would recombine with the protons to form again hydrogen atoms. However, applying an electric field the electrons or protons are pulled outside the chamber, and may be injected into the accelerator. Of course, electrons can be obtained even more easily, by heating some object (the cathode) like in your TV tube.

The first accelerators were linear accelerators. The protons or electrons were accelerated by means of a very strong electric field. The main objective was to obtain as high a voltage as possible. The best known machines in that genre are the van der Graaf machine and the Cockcroft-Walton accelerator. The maximum that can be obtained this way is of the order of 15 million Volts, i.e. a particle may be accelerated to an energy of about 15 MeV. That is not much on the scale of things, since not many particles have a mass less than 15 MeV, in fact apart from the zero mass particles there is only the electron. These accelerators were used to study nuclear physics, but they played no role in the discovery of new particles.

The next type of accelerator is the cyclotron, constructed first by Lawrence in 1930 at Berkeley. This machine consists of a round box cut in two, with a magnetic field perpendicular to it. An electric field voltage is maintained between the two halves. Protons are injected in the middle of the box, and start circulating: the magnetic field curves their path. Cleverly switching the polarity of the electric voltage when the protons are completely in one of the two halves it may be arranged that the voltage is such as to accelerate the protons when moving from one half of the box to the other. As the protons accelerate they make larger and larger circles until finally they are extracted from the machine. During the whole process the magnetic field is kept fixed.

Ernest Lawrence (1901–1958), Nobel prize 1939. He invented the cyclotron in 1929, and in the first picture he holds the first working model in his hands. He constructed it in the fall of 1930 together with his PhD student Milton Stanley Livingston (1905–1986). The second photograph shows the model in detail. It accelerated particles to 80 keV (0.08 MeV). The cyclotron developed in time to larger and larger size, and after World War II, when ample funds ware made available to nuclear physics, the development continued at a strong pace. It culminated in the US in the machine at Fermilab near Chicago, accelerating protons to 1 TeV (1 000 000 MeV).

Most important, Lawrence generated a whole group of machine builders, creating the machines at Fermilab, Brookhaven, SLAC (near Stanford) and Argonne (near Chicago). Lawrence himself was preceded by Rolf Wideröe (1902) from Norway. Wideröe has been called the first designer of accelerators; he constructed machines himself and inspired Lawrence and also Touschek who developed the first electron–positron collider.

The crucial idea behind the cyclotron is that the revolution time of the particles does not depend on the velocity. That is, at low energy they circulate in a small orbit, while near the end they move much faster, orbiting at a larger circle. The revolution time however remains the same. It is this principle, first seen experimentally and quickly understood theoretically that made the cyclotron such a viable machine. The cyclotron principle has been termed to be the single most important invention in the history of accelerators.

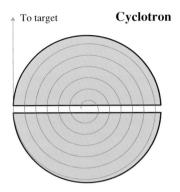

The figure above shows the path of the protons while accelerating. The magnetic field is perpendicular to the paper.

This device is in a sense the grand daddy of all circular accelerators. A particle is made to go around, its path being curved by a magnetic field. There is repeated acceleration at certain positions. For the cyclotron depicted here acceleration takes place when the protons cross the gap between the two boxes.

The maximal energy that can be reached with a cyclotron is about 1 GeV (= 1000 MeV). The machine will then fit into a box of about 4 × 4 × 4 m. Higher energy requires sizes larger than practical. It becomes difficult to make magnets of the required size and strength.

To accelerate to energies above 1 GeV the particles are kept in a fixed orbit with magnets stationed at that orbit, but now the magnetic field is synchronously increased as the energy of the particles increases. In this way one can do with a narrow pipe which then contains the circulating beam. Magnets are placed all along the beam pipe, and at certain positions there will be acceleration by means of electric fields. A number of very clever inventions was needed to make this happen: it is not easy to keep a large number of particles in tight bunches and accelerate them to high energies. Special methods are needed to prevent the slow ones from falling back, or the fast ones from running ahead, and also sidewise the particles must be contained. The important step here was what is called strong focussing: design magnetic fields in such a way that the particles stay within the beam pipe. These machines are called synchrotrons. By 1960 there existed two large proton synchrotron machines: the AGS (Alternating Gradient Synchrotron) at Brookhaven (Long Island, near New York) and the PS (Proton Synchrotron) at CERN, Geneva, Switzerland. The energies reached were 30 and 28 GeV respectively, which is about 25 times the energy of the largest cyclotron. The diameter of these machines is about 200 m.

Proton Synchrotron

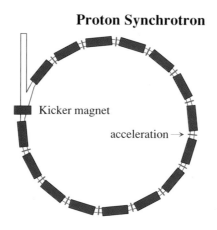

The picture above shows a sketch of a synchrotron. There are many magnets, placed along the ring. In the gaps between them (straight sections) acceleration is achieved by applying electric fields. The electric currents through the magnets are increased as the energy of the particles increases. When the top energy is reached the kicker magnet is activated to extract the particles from the ring towards experimental setups.

The synchrotrons, becoming bigger and bigger, have been under development till today. They will be reviewed in a separate section below.

Usually the particles accelerated in these synchrotrons are protons. The reason is that lighter particles such as electrons

Bruno Touschek (1921–1978). Touschek, born in Austria, did his thesis under the guidance of Heisenberg, and barely survived the war. His mother being Jewish, he was under constant threat of being arrested by the Nazis. With the help of several physicists he managed to escape for some time, but in 1945 he was arrested by the Gestapo and put into prison. He was frequently visited by Rolf Wideröe and they talked often of a new machine called the betatron. Around the end of February 1945 he was to be transported to a concentration camp near Kiel. Marching with a heavy pack of books and being quite ill, he fell on the ground somewhere in the outskirts of Hamburg. An SS officer took out his pistol and shot at his head, leaving him wounded and bloody in the gutter. As it happened he was only wounded behind the left ear. He made it into a hospital but was returned to a prison where he was liberated by the English.

After working at various places, notably Glasgow, he finally settled in Italy. In 1960 he proposed the first electron–positron collider where electrons and positrons move in the opposite direction in the same magnetic ring. Actually, the idea of a collider was due to Rolf Wideröe as early as 1943. Wideröe even patented the idea. It is, however, Touschek who came up with the idea of using one ring for two beams, and he constructed a machine (AdA) along these lines.

Unfortunately Touschek was addicted to both smoking and alcohol, and he died prematurely in 1978, in Switzerland. With him physics lost a great man whose impact was tremendous.

See also: *The Bruno Touschek Legacy*, CERN 81-19 yellow report.

suffer much higher losses when bent by a magnetic field. When the path of a charged particle differs from a straight line it radiates; the most elementary form of this effect is an antenna, where electrons are oscillated in the antenna and then radiate radio waves. A particle made to follow a circular orbit emits radiation, called synchrotron radiation, and that implies a loss of energy and thus de-acceleration. An electron of a given energy in a circular machine emits much more radiation than a proton of the same energy. The reason is that the amount of radiation depends on the velocity of the particle, not on its energy. An electron of the same energy as some proton has a velocity much closer to the speed of light than that proton which is 1800 times as heavy. Anyway, the consequence of this is that with a machine of given radius protons can be accelerated to much higher energies than electrons. On the other hand, the physics of proton collisions is very different from that of electron collisions, and that may be a deciding factor.

Another fundamental breakthrough occurred in a laboratory in Frascati, Italy. Bruno Touschek succeeded in 1961 for the first time to make a collider with a single storage ring. In such a collider two beams of oppositely charged particles circulate in opposite directions. At certain intersections the beams can be made to collide. The collision of a particle in one of the beams with a particle in the other beam produces an energy bubble that has twice the energy of each of the particles. In stationary target machines the energy of the bubble obtained when colliding an accelerated particle with a particle at rest is actually considerably below the energy of the particle in the beam, simply because the target particle recoils and a lot of energy goes uselessly into the movement of the energy bubble (kinetic energy of the bubble) after the collision rather than in the bubble itself. For example, colliding a 900 GeV particle with a proton (proton mass = 0.938 GeV ≈ 1 GeV) at rest produces a bubble of only 41 GeV,[d] moving however with a speed such that the kinetic energy of the bubble is

---

[d]Here is the equation: $41 \approx \sqrt{2 \times 900 \times 0.938}$.

859 GeV. In a collider the bubble resulting after a collision is at rest, and no energy is lost to the movement of the bubble as a whole. All of it is available for the production of new particles.

There is of course a big disadvantage to colliders: one has very little choice concerning the target material. Furthermore, the probability of a collision is quite limited, because the beams must be considered rather gaseous, from a particle point of view, and colliding beams tend to pass through each other without much action. The great problem with colliders is to get high luminosity, that is many particles in the beams and the beams as concentrated as possible. It is like colliding two needles, points ahead. The older machines (stationary target machines) used a target at rest made from ordinary material, which has many, many particles per unit of volume, and thus could be made sufficiently large to absorb the whole beam. Also, with colliders, there can be no high-intensity secondary beam configurations (see below). All in all, things like neutrino experiments which require a secondary beam (neutrinos) of very high intensity cannot be done with colliders.

There are three kinds of colliders: single-ring colliders, intersecting ring colliders, and linear colliders.

In a single-ring collider one injects particles in one direction, and the antiparticles in the opposite direction. For example one may use electrons and positrons. The same electric and magnetic fields can be used to accelerate the particles, because for a given electric field electrons and positrons are accelerated in the opposite direction, and they bend in the opposite way in the same magnetic field. There are some complications here: the beams of particles and antiparticles must be kept separated, crossing only at well-defined intersection points. Positrons are easy to make through pair production, and in fact the first such machine (Touschek's machine, called AdA, Anello di Accumulazione), 250 MeV per beam, ready in 1961, was an electron–positron machine.

The intersecting ring colliders have two rings that need not contain particles of opposite charge. Then one can use protons in both rings and they can be accelerated to much higher energies than electrons. At this time the maximum energy obtained for electron–positron collisions is 209 GeV (collision of two beams of 104.5 GeV each), while for proton collisions the energy is around 2000 GeV (at Fermilab). The 200 GeV electron machine called LEP (Large Electron Positron collider) is located at CERN.

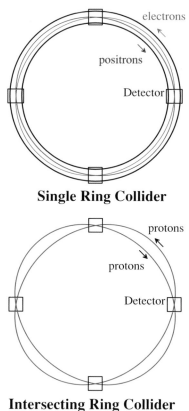

Single Ring Collider

Intersecting Ring Collider

It should perhaps be mentioned that it is nowadays possible to produce antiprotons in sufficient quantity needed for collider purposes. Some very clever inventions were required to achieve this; in 1984 the Nobel prize was awarded to C. Rubbia and S. van der Meer, and part of the reason was precisely this development. The

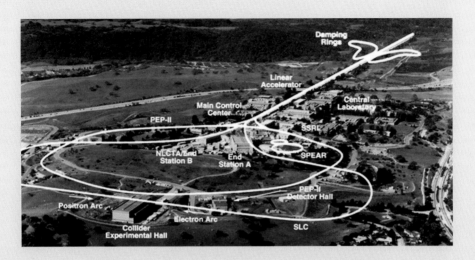

Aerial view of SLAC at Stanford near San Francisco.

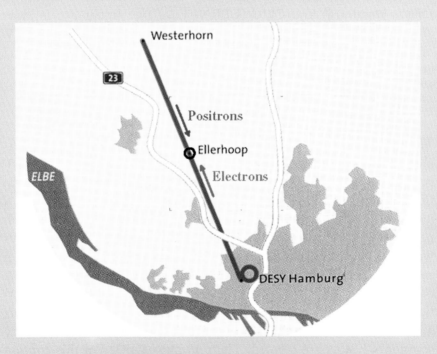

Artist view of TESLA. The machine will have two linear accelerators, each with a length of 15 km. The energy will be 500–800 GeV.

physics here was the discovery of the W and Z vector bosons, using proton–antiproton collisions.

Another breed of machines are the linear colliders. They essentially consist of two linear accelerators shooting against each other. The existing or proposed machines are exclusively electron–positron machines; there are no significant losses due to synchrotron radiation because the particle trajectories are not bent. As all acceleration must be achieved in one go, such machines tend to be very long. The most important machine of this type is located at SLAC (Stanford Linear Accelerator Centre). It is not strictly without bending; the same linear accelerator is used for both positrons and electrons and then they are bent around a half circle (one to the left, the other to the right) to collide head on. They have added single ring electron–positron colliders: SPEAR and PEP.

The planned (not yet approved) linear colliders TESLA (at DESY, Hamburg) and CLIC (CERN) have two linear accelerators and will produce collisions of 500–800 GeV and 1000–3000 GeV. The beams must be exceedingly small (well below one micrometer) and aimed very accurately onto each other. The mind boggles!

6.3 Secondary Beams

The number of protons in a stationary target machine is considerable. For example, in 1963, the CERN PS, operating at 25 GeV produced about 10^{12} high energy protons every 3 seconds. If in the collision with a target a particular particle is produced with a probability of 1 in a million, one still obtains a million of such particles every three seconds. A standard technique in those days was to aim the protons from the machine at some target (often made of metals such as beryllium, copper or tungsten). In the collision of the protons with the nuclei in the target many particles are created, and one can then set up a system of magnets and shielding such that certain particles are selected. In this way secondary beams are created. The relevant particles in this context are pions and kaons. The various particles will be discussed elsewhere, but some of their properties will be mentioned here. You will not find the pions (π-mesons) or kaons (K-mesons) in the list of elementary particles, because they are bound states of a quark and an antiquark. The following table shows the configurations.

name	symbol	quark content		charge	mass (MeV)
pos. pion	π^+	u	$\bar{d}$	$+1$	139.57
neutral pion	π^0	$u\bar{u}$ and $d\bar{d}$ mix		0	134.98
neg. pion	π^-	d	$\bar{u}$	-1	139.57
pos. kaon	K^+	u	$\bar{s}$	$+1$	493.68
neutral kaon	K^0	d	$\bar{s}$	0	497.67
neutral antikaon	$\bar{K}^0$	s	$\bar{d}$	0	497.67
neg. kaon	K^-	s	$\bar{u}$	$+1$	493.68

As the reader can see, the π^+ and the π^- are each other's antiparticle. Similarly K^+ and K^-, and K^0 and $\bar{K}^0$. The masses of a particle and of its antiparticle are always the same. The π^0 is its own antiparticle.

We have not indicated the color of the quarks; that is always a combination of color and the corresponding anti-color. For example, a positive pion is a mixture of bound states of an u_r and an $\bar{d}_r$, an u_g and an $\bar{d}_g$ and an u_b and an $\bar{d}_b$.

Pions are relatively light particles with a mass around 135 MeV (again, the electron has a mass of 0.511 MeV, the proton 938 MeV). They have baryon number zero, thus the proton cannot decay into pions. There are three types of pions, differing in charge but with approximately the same mass. The π^0 decays very fast, too fast to make secondary beams, but the charged pions have a relatively long lifetime. The π^+ decays for the most part into a positive muon (μ^+) and a muon-neutrino, and it decays in about two hundreds of a micro second. That is enough to actually go quite some distance, say a hundred meters. That depends of course on the energy of the pion. If there was no relativistic effect the pion, moving with a speed very near to the speed of light $(3 \times 10^{10}$ cm/s), would cover 5.2 meters before decaying. The theory of relativity tells us that there is an increase of the lifetime when observing a fast moving particle. A 1.3 GeV pion has an energy of ten times its rest mass, and the time dilatation effect is by that same factor of ten. A pion of this energy will therefore on the average travel 52 meters.

K-mesons come in four varieties. The charged kaons have a lifetime about half that of the charged pion, namely 1.23×10^{-8} seconds. The situation with the neutral kaons $(K^0$ and $\overline{K^0})$ is quite complicated because mixing between the two types plays an important role. The upshot is that there is a short lived kaon, notation K_S, with a lifetime about 300 times shorter than that of the charged pion, and a long lived kaon K_L with a lifetime that is twice that of the charged pion. So, the K^+, K^- and K_L are evidently suitable for secondary beams. All of these particles are created in abundance in proton-target collisions (many more pions than kaons as the pions are lighter), and it is quite possible to construct beams containing such particles with a reasonably well defined energy.

The bending of the trajectories of charged particles in a magnetic field depends on their momentum: the higher the momentum (and thus also the energy) the smaller the bending. One can there-

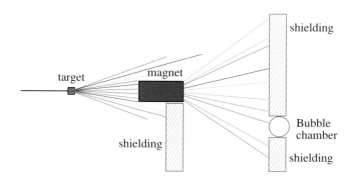

fore use magnets to select charged pions or kaons of a certain energy. The experimental apparatus would be installed at a certain angle from the magnet, and thus receive only particles of the corresponding energy. The particles going in other directions would be caught in the shielding. The picture shows a very simplified drawing of the set-up.

In the sixties many bubble chamber experiments were done using kaons. A typical experiment would, for example, take 100 000 to several millions of pictures of charged kaons with an energy of 5 GeV interacting in a bubble chamber. In these reactions many new unstable particles were created, with masses from 500 MeV upwards. It took quite some time and effort before the systematics of all these particles was unraveled, but that is history now. At this time they have all been understood as certain bound states with quarks and antiquarks as basic constituents, quite like the pions and kaons themselves, or protons and neutrons.

Neutrino beams may also be constructed, but they are more like tertiary beams. Neutrinos are essentially not produced directly in proton-target collisions, but the abundantly produced pions and kaons decay rather quickly into something usually including neutrinos, and that yields enough neutrinos to do experiments with. Neutrinos have no charge, and hence one cannot use magnets to select neutrinos of a certain momentum.

6.4 The Machine Builders

If there is one group of people that has made all progress possible, it is the group of laboratory directors, accelerator engineers and applied physicists. They constructed the instruments, the accelerators, that make experimentation possible. Often they are not in the limelight, as a rule the man at the end of the line doing the actual discovery receives the main attention.

Particle physics is very much driven by technology. Almost any advance in technology translates directly into improved accelerators, and thus to further experimentation. Experimentation today

Wolfgang (Pief) Panofsky (1919). Pief created the Stanford Linear Accelerator Centre, and developed a linear electron (and positron) accelerator. SLAC is arguably the most successful particle physics laboratory, generating three Nobel prizes. He was SLAC director from 1961 till 1984. I was there in 1963 during SLAC's building phase, and I was deeply impressed by Pief's leadership, knowledge and intelligence.

Educated in Princeton and Caltech (PhD), he participated in the Manhattan Project (atomic bomb) and after a period in Berkeley joined the Stanford University faculty in 1951. He very much involved himself in arms control issues, and remains an important US government advisor to this day.

His achievements are immense, and he received a large number of distinctions. Ironically not the Nobel prize.

His father, Erwin Panofsky (1892–1968), was a most famous German art historian. Being Jewish, he fled Nazi Germany in 1934 and after a short while went to the Institute for Advanced studies in Princeton. He had another son, Hans Panofsky (1917–1988), also very intelligent, who advanced the understanding of clear-air turbulence and the dispersion of pollutants. When both sons studied at Princeton University their intelligence was quickly recognized; as one of them appeared slightly smarter than the other they were dubbed the smart and the dumb Panofsky.

Pief, being in Munich, was once asked if he wanted to go to some museum. He answered: my father often spent hours explaining pictures to me and at some point I decided not to see any more of them.

Robert Wilson (1914–2000) and **John Adams** (1920–1984). In 1967 Robert Wilson, up to then working at Cornell University, started building Fermilab near Chicago. That included designing buildings, such as the Fermilab main building (the Hi-rise) as well as making various sculptures distributed over the Fermilab site. The Fermilab accelerator was made under budget. He was director till 1978.

Wilson, born in Wyoming, was a most remarkable man. He participated in the atomic bomb project despite his pacifistic views. Later, like Panofsky, he involved himself in anti-war activities. Ben Lee (heading the Fermilab theory group) told me that in 1971 after the Amsterdam conference (where 't Hooft reported on the renormalizability of gauge theories) Wilson actually urged him to work on those theories.

Lab directors cannot possibly be nice to everybody. In Fermilab's building phase Wilson's attitude was famous: he would fire anybody not busy working. At one time such a victim responded: you cannot fire me. Wilson asked why not, to which the man answered that he actually did not work for Fermilab.

Adams did build two accelerators at CERN: the PS (1953–1959) and the SPS (1971–1975). He also served as CERN's director from 1971–1980. He had no formal qualifications, but that did not keep him from being an extraordinary designer and engineer. He had a keen sense of competition with Wilson, his US counterpart at Fermilab. When Fermilab decided to build the Tevatron (using superconducting magnets) he expressed to me the desire to do the same at CERN, only better. This was not to be; CERN decided to build LEP. The decision was taken when Adams was director; despite his own preferences he, in the end, supported LEP.

depends on micro-electronics, solid state physics, low temperature physics and the associated technical developments. No wire chambers could have been possible without the incredible advances in electronics of the last 50 years. But perhaps the most dramatic progress has been in accelerator development. Let us take a look at what happened. We shall skip cyclotrons and start with the development of what is called fixed target accelerators, mainly accelerators where a beam of protons is accelerated, extracted and sent into a target. As described before, much of the energy is lost in the form of kinetic energy of the resulting bubble, and to be fair one should for such machines list the energy that goes into the energy bubble itself. For example, the CERN PS (the first big accelerator at CERN) produces protons of about 25 GeV, but after the collision of a proton in the beam with a proton in the target one gets an energy bubble of only 7.3 GeV, the rest of the energy being in the forward movement of the bubble. In colliders, where two particles of the same energy meet head on, the resulting energy bubble is at rest and all energy is in that bubble. Physicists speak of the center of mass energy when talking about the energy of the bubble.

The first big fixed target machine, called grandiosely the Cosmotron, started operation in Brookhaven around 1952. The energy reached for the proton beam was a bit above 1 GeV. Around 1957 the Russians came with a machine of 7 GeV (bubble energy 3.8 GeV) located in Dubna near Moscow. In 1959 CERN started operation of the PS, proton synchrotron, with an initial energy of 23 GeV (7.1 GeV energy bubble). Slightly later the Americans completed the AGS machine (Alternating Gradient Synchrotron) at Brookhaven, Long Island, producing 30 GeV protons (7.6 GeV energy bubble). Both machines were built on the principle of a circular pipe of fixed dimensions with continuously increasing magnetic field and with what was called strong focussing to keep the particles inside the pipe. The method of strong focussing, producing narrow stable beams, was re-invented by a group of physicists in Brookhaven.[e] It was crucial for the development of these machines.

The next step was achieved mainly by making the machines larger. While the CERN and Brookhaven machines had a diameter of about 200 meters, the next generation of machines had a diameter of about 2 km. One such machine was built at CERN (the SPS, Super Proton Synchrotron) and an energy of 400 GeV (energy bubble of 26 GeV) was achieved. The machine at the Fermi National Laboratory (FNAL, near Chicago) produced protons of a slightly higher energy, namely 500 GeV (30.5 GeV energy bubble). Using superconducting magnets they later achieved doubling of the energy, to 1000 GeV = 1 TeV. That machine is called the Tevatron. There are about 900 superconducting magnets along the ring. Iron core magnets can have a magnetic field up to about 2 Tesla, superconducting magnets up to 10 Tesla. Superconducting magnets have no iron core, but the currents through the magnet coil are exceedingly large, as much as 5000 Ampere. For such currents any resistance of the wire of the coils is fatal, and one uses for these wires superconducting material (some niobium alloy). The superconducting coils, at least when kept at about 270 degrees below zero (the temperature of liquid helium) have exactly zero resistance to the electric currents. Superconducting magnets are dangerous devices: if through some accident or error somewhere a little part of the coil heats up by a few degrees the superconducting property is lost and the large current generates much heat at that spot. This then heats up the surroundings, etc. The consequences are disastrous. It is an almost explosive happening, and a number of special measures have to be taken, among others to capture the helium which becomes gaseous the moment the temperature rises. The technology at these accelerators is truly fearsome!

---

[e]The first inventor was Nicholas Christofilos, a Greek engineer. He patented it, but unfortunately his work was only published in preprint form and therefore not well known.

It is interesting to plot the bubble energy (the center of mass energy) achieved at these machines against the year in which they were constructed. Amazingly a straight line results if for the y-axis a logarithmic scale is used. Normally, plots are linear, that is every mark on the scale is a fixed amount above the previous mark. Thus then the scale reads sequentially, 1, 2, 3, 4, etc. On a logarithmic scale each mark is a fixed factor above the previous one, the scale is then 1, 10, 100, 1000, etc. That is the same type of scale that you can find on old-fashioned slide rules. That scale applies in many cases, such as the development of the world population. This accelerator plot is called a Livingston plot, after the American machine builder. The plot (see figure below) does not show all accelerators ever built but mainly the top achievers.

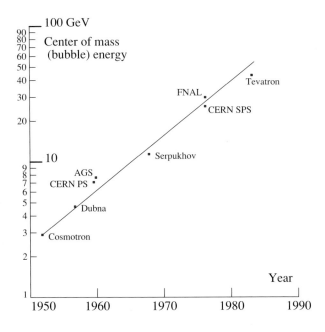

A similar plot can be made for machines with storage rings (called colliders): proton–proton, proton–antiproton, electron–positron, electron–proton and positron–proton colliders. Proton–

proton and electron–proton or positron–proton colliders require two intersecting rings, the particle–antiparticle colliders can do with one ring in which both particles and antiparticles circulate in opposite directions.

Many other machines followed after AdA (the machine made by Touschek), with the largest one being LEP at CERN, design energy 91 GeV per beam (a 182 GeV energy bubble). In the final months of LEP's existence, engineers in a splendid demonstration of their prowess drove the LEP center of mass energy to 209 GeV. In the figure below the Livingston plot for colliders is shown. Again, not all existing or past machines have been included.

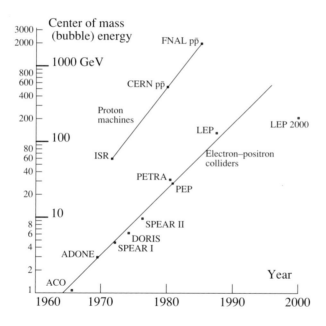

ACO (see subject index for machine acronyms) is a machine at Orsay, France. Like most electron-positron colliders it is a single ring machine. ADONE is the successor to AdA, located in Frascati, Italy. PEP, SPEAR I and its upgraded version SPEAR II are at SLAC (Stanford Linear Accelerator Centre) at Stanford, California.

DORIS and PETRA are at the DESY laboratory (Deutsches Elektronen Synchrotron), Hamburg, Germany. That laboratory also has the only electron–proton collider in existence, called HERA, with a bubble energy of 400 GeV. LEP and ISR are at CERN. The ISR is a proton–proton collider with two rings. The CERN and FNAL (Fermilab near Chicago) $p\bar{p}$ colliders use the SPS and Tevatron rings, containing a proton as well as an anti-proton beam. Not on the plot is the LHC, a proton–proton machine being built in the LEP tunnel at CERN. If all goes well it will produce energy bubbles of 14 TeV = 14000 GeV in 2005. It will probably be the last proton machine to be built. The LHC has two beams of maximally 7 TeV each, crossing each other at four points (intersection regions) where detectors will be mounted. Think of it: the circumference of this machine is 26.67 km, thus one will need about 25 km of superconducting magnets. And each of these magnets is a technological tour de force! If you want to know more about it visit CERN's website (www.cern.ch). Incidentally, CERN is the place where the World Wide Web was invented, by Robert Cailliau and Tim Berners-Lee. Its aim at that time was to facilitate the collaboration of many experimental groups at various universities mounting experiments at CERN.

An American super machine (the SSC), a proton-proton collider, was terminated prematurely. The 11 billion dollar machine would have achieved a center of mass energy of 40 TeV. Now there is an empty underground tunnel of 48 km. On the drawing board now are linear colliders, where beams produced in a pair of very long accelerators are made to collide head on. The designs are breathtaking, really incredible. There is also talk about muon–antimuon colliders, but these are difficult machines.

7

The CERN Neutrino Experiment

7.1 Introduction

In 1959 the proton synchrotron at CERN, Geneva, Switzerland started running, followed a little later by the AGS (Alternating Gradient Synchrotron) at Brookhaven, Long Island, USA. Both machines accelerated protons to approximately the same energy, 28 and 30 GeV respectively. They were the biggest machines of that time, with a diameter of about 200 m. The intensity was respectable: about 10^{11} protons per 3 seconds. Physicists started working with these exciting new toys. The era of big high-energy physics had started.

At that time the state of affairs concerning particle physics in Europe (with England as an exception) was simply dismal. Ravaged after World War II, Europe started to get back on its feet. Perhaps the biggest problem was the absence of leading physicists; many Jewish physicists had left for the US, and notably also E. Fermi (who was not Jewish, but his wife was). No substantial experimental effort existed anywhere in Europe before 1957, although here and there cyclotrons were built, used however almost exclusively for nuclear physics (the study of the structure of the atomic nucleus). In 1957 a 3 GeV proton synchrotron called Saturne started up in France, but it did not play any role of significance in the development of particle physics that I know of, except perhaps in educating experimenters.

In the US the influence of Fermi cannot be overestimated. To me he is an example of how one man can make a big difference.

Bruno Pontecorvo (1913–1993) and **Melvin Schwartz** (1932). Pontecorvo has had essentially all the ideas for neutrino experiments. He was the first to think of the so-called chlorine-argon method for detecting neutrinos (including neutrinos from the sun), and he also introduced neutrino mixing (in 1957). The chlorine-argon method was put into practice and further developed by Davis, who demonstrated that reactor antineutrinos were different from neutrinos, and who detected neutrinos from the sun (Nobel prize 2002).

The idea for neutrino experiments at the big machines is due to both Schwartz and Pontecorvo. Schwartz went on to do the experiment, together with Lederman, Steinberger, Goulianos, Gaillard, Mistry and Danby. Lederman, Schwartz and Steinberger received the 1988 Physics Nobel prize for this landmark experiment.

Pontecorvo, a devoted communist, already politically active in the thirties, moved to Russia in 1950 in a somewhat fugitive way. He was one of those scientists who were blamed for defecting to Russia taking along atomic bomb secrets. In his case there is not much substance to that; he was never actually involved in weapons research. He just believed in communism. I guess he paid the price.

Schwartz later suggested beam dump experiments at SLAC, and he had in fact a short run. He showed me a few pictures (dubbed Melons by some) at the time of the 1971 Amsterdam conference, and to me it was immediately clear that he had observed neutral current neutrino events. Conflicts with the SLAC directorate (Panofsky) led a somewhat embittered Schwartz to leave physics, and he started a successful electronics company called Digital Pathways. Personally I believe that he was a better physicist than businessman. He is too honest.

Tsung Dao Lee (1926) and **Chen Ning Yang** (1922). They shared the 1959 Nobel prize for their work on parity violation. This concerns the behaviour of physics laws when considered through a mirror. Thus do two sets of experiments, and observe the results directly, but also, independently, in a mirror. The question is whether the laws deduced from such experiments will be the same. They analyzed the situation assuming that this is not so, and indeed it is not. When observing the decay of a pion at rest into a muon (and an antineutrino) the muon spins in a left-handed way along the direction of movement in ordinary space, while in the mirror one observes a muon spinning in the opposite way.

Lee and Yang collaborated till 1962, when they broke apart for reasons of their own. In my opinion the sum was better than the two individually, an example of synergy. They had just started on a systematic investigation of vector bosons (the W and Z of weak interactions), and there is no telling how far they could have gone in developing the Standard Model. Lee was very strong on Feynman diagrams, while Yang was together with Mills the originator of gauge theories (also called Yang-Mills theories) that are an essential ingredient of the Standard Model.

The idea of Schwartz for a neutrino experiment caused Lee and Yang to analyze the situation in precise detail. Their work, published in 1960, became the guiding light for both the Columbia and CERN neutrino physicists. Together with Markstein from IBM Lee and Yang did one of the first large scale computer calculations concerning the possible detection of the vector bosons in a neutrino experiment. None were actually seen, they were too heavy.

In Europe the problem was different from country to country. In the Netherlands there were essentially no experimenters in the domain of particle physics, and in fact there was simply no money available to start anything substantial. As to theorists, there were a few of international stature, in particular H. Kramers who contributed in a fundamental way to particle theory. But after Kramer's death in 1952 only one of his pupils (N. van Kampen) worked in a prominent way in particle physics (to which he contributed substantially), switching however to statistical mechanics around 1958.

In most other European countries the situation was equally sad. L. de Broglie, in France, had a very negative influence on the development of theoretical physics there. In Germany there was an aging W. Heisenberg whose image had been tarnished, and who moreover had gone off on a tangent theory-wise. There were some excellent theorists there, but they worked mainly on rather highly abstract subjects, far away from experiment. Italy and England were in better shape, especially in the field of particle theory but also experimentally. In Switzerland there were a number of excellent theorists, notably Pauli and Stückelberg.

In Europe the big breakthrough was the creation of a new international organization for doing research. The CERN treaty was signed in 1953, and only seven years later the first big machine started up in Geneva. Since then CERN has been essential to particle physics in Europe, not only because of the big machines, but also because it functions as a centre of physics that no country could afford by itself. There you could meet all the well-known physicists, mainly the Americans, that had made and did make the field.

This is not the place to start describing the European state of affairs after World War II, although I am not clear who else is doing it. Usually they paper it over. I entered the domain as late as 1961, and cannot competently speak on these matters. When I arrived at CERN in 1961 however, there was no question about it: in particle theory we were nowhere comparable to the US. I myself did not know the difference between kaons or pions at that time, although they had been around since the war. At CERN, it was an American physicist, S. Berman (a student of Feynman), who did put me on the right track, neutrino physics. There I also started a long-time collaboration and friendship with John S. Bell, who was one of the very rare people at CERN actively interested in weak interaction theory and neutrino physics.

I started by trying to compute the production of vector bosons by neutrinos. If the vector bosons had a mass of less than 1.5 GeV its detection would have been within experimental reach at that time. The reader knows already that they are much heavier, around 80 GeV, but we did not have any idea about that. These calculations were quite tedious, had been done before by T. D. Lee, P. Markstein and C. N. Yang, and were essential to the experimenters who were starting up the CERN neutrino experiment. It is in this way that I came into contact with that group, at the end of 1962. The experiment started in June 1963, and from the point of view of physics it was to me a happening of overwhelming influence. Nothing compares to entering a new domain where no one has been before, and that was what I experienced by watching the experimental results come in. In the end the experiment was a failure, but that does not take away my feeling towards that experience.

If asked why that experiment was a failure I would say that it was simply because this was still Europe getting back on its feet. We were all learning. Hardly anybody had any experience to speak of. Others, in particular Lee and Yang, told us what to look for. Technically speaking the experiment was a great success. The CERN engineering staff, among them a sizeable number of Dutch engineers, was second to none. The weak point was physics, and that included the theoretical part. The responsibility for the theoretical part, if there was really anything like that, was mainly in the hands of J. S. Bell and myself, and to this day D. Perkins, experimenter from England, blames us for not discovering scaling (never mind what it is) in the plots that

Leon Lederman (1922) and **Jack Steinberger** (1921). They shared with Schwartz the 1988 Nobel prize for the discovery of the muon-neutrino at the Brookhaven neutrino experiment.

After the original idea of Schwartz two groups (BNL and CERN) were formed to do the actual experiment. CERN had the advantage. Steinberger, to the dismay of the others in the BNL group, took a sabbatical and joined the CERN group. However, the CERN experiment was aborted unexpectedly when von Dardel, a Swedish physicist, discovered errors in the event rate estimate. The CERN group restarted, using beam extraction and adding van der Meer's focussing horn (horn of plenty), but they lost their advantage. As Schwartz put it: "Early in 1961 it looked like Jack and his associates at CERN would certainly have the first neutrino events. Then we received a piece of fabulous news. CERN had cancelled the neutrino experiment."

Steinberger returned to Columbia, and re-joined the BNL group although not without some arguments. A few years later he went again to CERN. He did several fine experiments, but he was also the *auctor intellectualis* of the split field magnet, a detector with a complex magnetic field that supposedly could unravel anything. However, the thing was too complex and became a computer programmer's nightmare, or rather cemetery. Cynics changed spl into sh.

In 1950 Steinberger discovered the neutral pion (with Panofsky and Steller). He also produced a theoretical explanation that became important later on.

In 1978 Lederman became the director of Fermilab, as successor to Wilson. Before, in 1977, he essentially discovered the bottom quark in a Fermilab experiment.

he produced on the basis of events found in this neutrino experiment. Scaling, incidentally, was discovered much later, at the end of the sixties, by the theorist J. Bjorken from Stanford, and the SLAC experimenters that did the relevant experiment received the 1990 Nobel prize for this work (not Bjorken). Well, sorry about that, Don.

Scaling was not the only thing missed in the CERN neutrino experiment. But let me discuss it systematically. However, first the experimental set-up must be described, and the physics objectives as seen at that time.

7.2 Experimental Set-up

The Italian physicist Pontecorvo (Joint Institute for Nuclear Research, Dubna near Moscow) together with the American physicist Schwartz (Columbia University) can be credited for suggesting experiments with high energy neutrino beams made at the big machines. The main idea of neutrino experiments in the form to be described was due to Schwartz. He started thinking about these things after stimulating conversations with T. D. Lee, who asked if it would be possible to do weak interaction experiments at high energy. The basic set-up suggested by Schwartz in October 1959 along the lines of his and Pontecorvo's ideas is this.

First let the accelerator run at the highest possible energy, and then collide the protons with some stationary target. In the proton-nucleus collisions in the target many, many charged pions will be created. These charged pions decay mainly into a muon and a neutrino, but you have to wait a bit because the decay occurs only after some time. Thus after the target there is a decay area of some 25 m in which the pions can decay. After that there comes a massive amount of shielding, meant to block every particle except neutrinos. The latter cannot be blocked within earthly distances. After the massive shielding then there are the detectors.

The big point about the idea is the flux of neutrinos that one obtains in this way. Are there enough to give a reasonable probability of a reaction in the detector? The idea, at that time, was outrageous. But as it happens the neutrino flux is adequate for experimental purposes, although only barely.

Much of the feasibility of the experiment depends on the size of the detector. Evidently, as the amount of matter in the detector increases, the number of neutrino induced events will go up proportionally. As luck had it, the Japanese physicist S. Fukui had just invented a new device, the spark chamber. The great thing about spark chambers is that one can make the plates of relatively thick and heavy material. In this way a detector of tens of tons could be constructed. Thus the detector is also the target. The suitability of using spark chambers for a neutrino experiment was first suggested by the American physicist Irwin Pless.

The earliest neutrino experiment was performed at the Brookhaven laboratory, by a group from Columbia University including Schwartz. This will be discussed later. For now the set-up of the CERN neutrino experiment will be described. The figure below gives a sketch of the experimental set-up at CERN in June 1963.

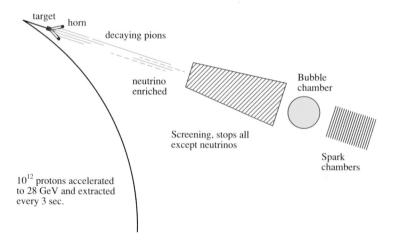

To increase the neutrino flux two important steps were taken. First, new at that time, full beam extraction was achieved. Before that one just placed a target inside the PS machine. Secondly,

Helmut Faissner (1928) and **Frank Krienen** (1917) talking to Yang at the time of the CERN neutrino experiment. Faissner and Krienen (and the original inventor, Fukui, who participated in the CERN experiment) were largely responsible for the spark chamber set-up. The quality of those chambers was excellent, and once the experiment started running they reproduced in a few days the BNL results (run of eight months, 56 events). Then the search for the *W* started in earnest, and in particular Faissner worked so hard at it that if it had been possible to create such events by sheer will power they would have been there. Alas, no such luck.

Later Faissner and his group in Aachen found the first electron neutral current event of the type neutrino + electron $\rightarrow$ neutrino + electron. That was in 1972, in a photo from the huge French bubble chamber Gargamelle exposed to a high flux neutrino beam. That was strong partial evidence in favour of what is now known as the Standard Model. It convinced many physicists of the correctness of gauge field theory applied to weak interactions.

Krienen was one of the really excellent Dutch engineers that started working at CERN right from the beginning. While there were virtually no Dutch experimental physicists at CERN, Dutch engineers such as Krienen, Kuiper, van der Meer, Middelkoop, de Raad, Zilverschoon and others had a major impact. As mentioned before, Krienen later developed digitized spark chambers. He contributed to many experiments, notably a big $g - 2$ experiment at CERN (this is discussed in Chapter 9) involving a muon storage ring. In 1982, retiring from CERN, he went to the US, and started a whole new career and a whole new family including two children.

Building the shielding for the CERN neutrino experiment.

Installing the Heavy Liquid Bubble Chamber.

The spark chambers with a track produced by a cosmic muon.

S. van der Meer invented a magnetic horn. This device, placed around the target, generated a magnetic field such that charged particles were focussed in the forward directions. These two measures led to an increase in the neutrino flux to the point that a bubble chamber could be used as detector: roughly one event per 2000 cycles for a bubble chamber of 0.75 ton. Since the bubble chamber is expanded and a picture taken at every cycle (3 sec.) of the proton synchrotron this means one event per 2000 pictures, or roughly 15 events per day if running optimally.

The spark chamber setup was quite elaborate. First there were relatively light spark chambers, supposedly showing finer details, which indeed they did. This region was called the production region. Following the production region there were two magnet coils around some 5 spark chambers. The idea was to determine the sign of the charge and the magnitude of the momentum of the charged particles coming out of the production region. Finally there was a set of quite heavy spark chambers, called range chamber, meant to determine whether the charged particles seen were muons or something else. This is based on the fact that muons can go a long distance through matter without doing anything, and only they could produce long tracks in the range chambers.

The experiment started in June 1963, and continued till the end of August. Results were presented at a conference at Brookhaven Laboratory by representatives of the engineering group (C. Ramm), the bubble chamber group (R. Voss) and the spark chamber group (H. Faissner). To me fell the task of presenting the conclusions. Quite an experience as you can imagine, since the "fine fleur" of the world's particle physics community was present. Moreover, this was in fact the entry of CERN into the big world of particle physics. The spectacle of a theorist presenting the conclusions of an experiment can perhaps best be appreciated by quoting V. L. Telegdi from the University of Chicago: "A theorist telling the experimenter what they are doing is like a newly married couple taking a gynecologist along for their wedding night."

7.3 Neutrino Physics

The theoretical side of neutrino physics was dominated by T. D. Lee and C. N. Yang. They had received the 1957 Nobel prize for their work on weak interaction theory, more precisely the analysis of parity violation in weak interactions, which is something that we need not go into right now. At Columbia University Lee inspired M. Schwartz, who devised the basic mechanism by which neutrino experiments were done. Lee and Yang wrote a paper investigating the physical aspects of neutrino physics, and this became the guiding light to the experimental groups.

The following two questions came to the foreground:

1. The two-neutrino hypothesis;
2. The vector boson hypothesis.

The neutrino hypothesis is the following. The neutrino made its entry in 1930, through a proposal by W. Pauli. Study of beta-decay (of which neutron decay is the prime example) showed that the total energy of the visible end-products did not add up to the initial energy. Here is neutron decay as understood now:

$$\text{neutron} \rightarrow \text{proton} + \text{electron} + \text{antineutrino}$$

Thus the energy of the proton and the electron did not add up to the energy in the initial stage, which is the mass-energy of the neutron, and we understand now that the remainder is carried off by the antineutrino.

This hypothesis was generally accepted. For a long time the neutrino remained a spooky particle, because it was only seen as an absence of energy and momentum. This changed in 1956, when F. Reines (Nobel prize 1995, shared with M. Perl for the discovery of the tau meson) and C. Cowan succeeded in observing neutrino induced events in scintillation detectors. The neutrinos came from the Savannah River nuclear reactor. A nuclear reactor produces a considerable flux of (anti)neutrinos due to beta decay of the fission products. The experiment solidified the neutrino idea, but it must be said that quantitatively speaking much was still unclear.

After the war other reactions were observed where presumably neutrinos carried off energy. Examples are:

pion → muon + neutrino

muon → electron + neutrino + antineutrino

For theoretical reasons people started to ask themselves if all these neutrinos were the same, i.e. if there was more than one kind of neutrino. More specifically, it was suspected that neutrinos (or antineutrinos) produced together with electrons might not be the same as those produced together with muons. Thus the neutrino (never mind that it is actually the antineutrino) in neutron decay, produced together with an electron, would be a neutrino of the electron type, while the neutrino from pion decay, produced together with a muon, would be a neutrino of the muon type. And in muon decay, with two neutrinos there would be one electron-neutrino and one muon-neutrino. This idea, that there are two kinds of neutrinos, is called the two-neutrino hypothesis. The way the idea is implemented is by means of two quantum numbers: electron number and muon number. Electrons and electron-neutrinos have electron number 1 and muon number zero, while negative muons and muon-neutrinos have muon number 1 but zero electron number. Of course, the antiparticles have the corresponding negative value for their muon or electron number. All other particles have zero electron or muon number. In pion decay one starts with muon number zero (the pion) and ends with a muon number zero (a positive muon has muon number −1, the muon-neutrino has +1). In muon decay, starting with a negative muon, the initial value of the muon number is +1. In the final state there is a muon-neutrino (muon number 1), an electron (electron number 1) and an anti-electron-neutrino (electron number −1).

All this amounts to the following. In a neutrino experiment the great majority of the neutrinos come from pion decay, and since

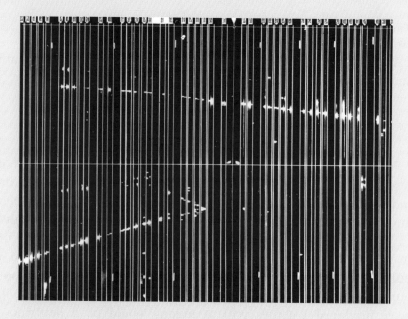

Here a special spark chamber picture from the CERN neutrino experiment. The neutrino beam was sufficiently intense to cause sometimes the occurrence of two events simultaneously. Both events show at least one straight ongoing track, typically a muon. It is from this type of event that the BNL group concluded to the existence of two neutrinos, as there always seemed to be a muon and practically never an electron. Electrons (and positrons) are very easy recognizable in a spark chamber because they produce a shower, a multitude of relatively small tracks. The CERN experiment produced in a short time many events as shown above, thus confirming the results of the BNL group and the existence of two neutrinos.

Neutral current events are characterized by the absence of either a muon or an electron. Just imagine that the muon is not there. The lower event would have been relatively easy to identify. The trouble is that stray neutrons coming somehow around the shielding could produce something quite similar. The problem of seeing neutral currents became one of how to eliminate that possibility. But as there was really no interest in neutral currents at that time no one thought of doing that.

Sometimes I try to imagine how history would have gone if indeed neutral currents had been established by the CERN experiment. I guess they would have become part of the weak interaction phenomenology, and that would be it. If there is no theoretical framework, which indeed there was virtually none at the time, it is very difficult to see that this points to some type of theory.

the pions decay almost exclusively into a muon and its neutrino it follows that the neutrinos in these experiments are almost exclusively muon-neutrinos. There are some electron-neutrinos in the beam, for example from kaons that decay into neutrinos and electrons (and other stuff), but they amount to very little.

The question is what these neutrinos do in the detectors. If the two neutrino hypothesis is correct then an event starting off with a muon-neutrino must wind up with either a muon or again a muon-neutrino in the final state. However, there should not be an electron in the final state, although electrons can appear in addition to the muon, see below. So here was experimental objective number one: establish if the two-neutrino hypothesis is correct. The procedure is simple: look to neutrino induced events and see if they have muons and/or electrons in the final state. The muon goes a long way unperturbed through material and thus produces long tracks, the electron gives rise to a shower, so this point can be easily checked provided you have neutrino events in the first place.

The vector boson hypothesis is another story. At this point there is no real need to delve into the theory of this object; the only thing to know is that they can be produced in neutrino events in addition to the muon. A vector boson may decay into pions or kaons, but in addition it will decay with some probability into an electron-neutrino and with the same probability into a muon-neutrino pair. This is the way a vector boson can be established: a muon-neutrino collides in the detector and produces a muon and a vector boson (and possibly some further debris, never mind that). Now the vector boson decays quite quickly and in a certain percentage of the cases it will go in a electron-neutrino or muon-neutrino pair. In those cases one observes in the final state with equal probability an electron (plus neutrino) or a muon (plus neutrino) coming from this vector boson. In short, since the neutrino is invisible one sees either two muons, or a muon and an electron. We did not make any effort to get the charges correct, or state precisely which is the particle, and

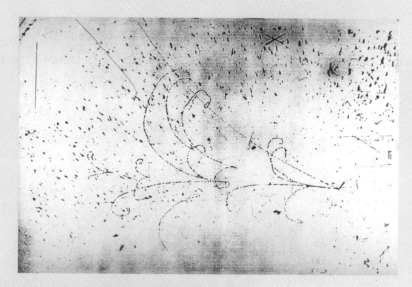

One of the very first bubble chamber pictures seemed to have just the right signature for a *W* production event. The small figure on the right shows the event redrawn. The neutrino beam entered from the right. After the collision several particles came out, and there was a recoiling nucleus (N in the drawing). The negative muon is recognized as such because it rarely interacts and does not lose energy. The remaining tracks are electrons and positrons

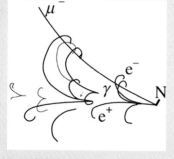

that lose energy fairly rapidly and they actually come to rest. The tracks curve due to a magnetic field. The positrons curve in the opposite way as compared to the electrons. What one sees is a shower. The photons, invisible, generated by a previous electron or positron, convert to electron-positron pairs (the seagulls). One such photon is indicated by a γ. There is a single electron (see arrow) perhaps kicked out of an atom.

In the first instance the shower was seen as due to a single positron. Thus it seemed as if there was just one muon and one positron. The *W* interpretation however was discarded later on because the energies and momenta did not check out. The angle between the muon and the positron was too large to be a *W* event. But in the first few days everyone thought we were going to get many such pictures, thus discovering the *W*. However, no such picture was ever produced again in the bubble chamber. It was a nasty little joke of Nature.

which the antiparticle, because at this point that does not matter. Just look for two muons, two long tracks, or one muon and an electron (or positron), i.e. one long track and a shower. That is the signature for a vector boson event.

In 1960 no one knew what the mass of the vector boson was, and the main issue then was whether neutrinos in the neutrino beam were sufficiently energetic to allow the production of such a boson. At CERN one started off with 25 GeV protons, but by the time one has neutrinos their energy goes down considerably. In practice vector bosons could have been observed provided they were lighter than 1.5 GeV. That is a far cry from the value established now: 80 GeV.

These were then the main objectives for the initial neutrino experiments. Two machines were, starting up, the CERN PS and the Brookhaven AGS. In fact, the Brookhaven machine would reach completion about 6 months after the CERN machine started, thus CERN had a 6 month advantage to achieve these physics goals. Since no one would see vector bosons in these experiments it became a matter of who would verify the two-neutrino hypothesis. That was where the prizes were, and indeed the 1988 Nobel prize was given for that. The history is amusing, so let's tell some about it.

7.4 The First Neutrino Experiments

After the initial idea of a neutrino experiment Schwartz undertook its implementation. He obtained money, collaborators, and most importantly, the cooperation of the director of the Brookhaven lab, the well-known physicist M. Goldhaber. While at CERN J. Adams, director till August 1961, was an engineer, Goldhaber was perhaps more aware of the potentialities and the importance of the experiment, and he was willing to grant the Columbia group the necessary privileges, including 8 months of running time. I guess he also risked his own life if I may believe Lederman's story as told in his book. Columbia physicists consider the Brookhaven machine their private property. US laboratories have usually experimental physicists as director, while at CERN one has had engineers and theorists next to some experimenters.

At CERN a neutrino group was formed under the direction of G. Bernardini, an Italian physicist. He had considerable experience in particle physics, and he was a first-class physicist. Most other members of the CERN group were newcomers to the profession. They had a head start, in principle, of six months on the Brookhaven group. There was naked competition.

So, the CERN group started out by designing a neutrino beam. For speeds sake beam extraction and the magnetic horn of van der Meer were not part of the first design. A target would simply be inserted into the proton machine itself. No kicker magnet, no beam extraction, no horn. Also no spark chambers, instead two bubble chambers and a large scintillation counter set-up with a Wilson cloud chamber. A group of three people computed the neutrino flux and the event rate on the basis of this crude set-up.

What happened then is very difficult to get straight as different people have different recollections, and I cannot claim to have the complete truth nailed down here. I arrived at CERN in September 1961, and did not know of all the commotion till long afterwards. Anyway, here is at least an important part of the story as I learned from various letters written in that period.

In May 1961 the physicist Guy von Dardel (from Sweden) discovered flaws in the neutrino flux calculations. Remarkably, he was not even a member of the CERN neutrino group, but had been asked to verify by measurement the estimates used for the initial pion flux. The synchrotron consists of a ring of magnets, separated by small straight sections. The target was to be placed in one of those straight sections. The pions coming out of the target would have to pass rather narrowly by the subsequent magnet in the ring, and the magnetic field influences the pions. The amount of influence is directly related to the magnitude of the straight section where the target would be put inside the machine. Von Dardel then investigated the original calculation and discovered

Helmut Faissner (1928), **Guy von Dardel** (1919) and **Giampietro Puppi** (1917) at the CERN terrace in June 1962. Guy von Dardel is related to Raoul Wallenberg, the Swedish diplomat who helped many Jews during World War II. He did that in Budapest, and at the end of the war, when the Russians entered Budapest they took him and he disappeared in the goelag, to die after two and a half years. For a long time the Wallenbergs, and notably also von Dardel tried to find Raoul, but until 2000 the Russians would not acknowledge his existence, or even that he ever was in Russia. On the web you can find out more about this.

Puppi, an Italian experimental physicist, got some fame for the introduction of the Puppi triangle, suggesting that among others the processes $\mu \rightarrow e + \nu + \nu$ and neutron $\rightarrow$ proton + e + ν would go at equal strength. That triangle was too simple and disappeared when Cabibbo introduced his angle (Chapter 3), but even so it contained an important truth: lepton-quark symmetry. It is that symmetry, often cited by Gell-Mann, that led Hara to introduce the fourth quark now called charmed quark (see at the end of Chapter 8).

Puppi was for a few years director of the research division of CERN. He was candidate for the position of director of CERN (after Weisskopf), together with Gregory, and we had a hard time deciding to whom we should be friendly. Gregory won.

The CERN terrace, where you can see the Mont Blanc on the horizon, is very popular among high-energy physicists. You can meet there just about everybody in the business. Many initiatives were started there, and many ideas were born in that environment. So far you can still smoke a cigar there.

that it contained errors on this point. Also, they had not taken into account a very elementary fact: when a pion decays the resulting neutrino does not go on in exactly the same direction as the pion, but it may deviate sideways, see figure below.

According to von Dardel it would take about 6 months to obtain some 3 or 4 neutrino events, with the machine totally dedicated to the neutrino experiment, suspending all other experiments. Von Dardel got very emotional about it and called it a scandal. In the summer of 1961 CERN decided to postpone the experiment.

At CERN the straight section had a length of 1.5 m. At Brookhaven it was actually 3 m, so the pions at Brookhaven were liable to be influenced a lot less than those at CERN. The Columbia experiment at Brookhaven started running in December 1961, and in 1962 the result was announced on the basis of 56 events obtained in 8 months of running time: there are two neutrinos. Mostly muons were observed, only a few electrons. The three leaders of the experimental group, Lederman, Schwartz and Steinberger shared the 1988 Nobel prize for this discovery.

Should CERN have gone ahead despite the set-back? No one knows for sure. But by all accounts the neutrino flux seemed too low, and the detector mass (no spark chambers) too small. One thing is certain: when Schwartz heard about the CERN decision to postpone he was overjoyed!

7.5 Vector Bosons

After the publication of the Columbia experiment results CERN found itself in the unenviable position of having been scooped. The consequence of that was clear: CERN would have to focus on the discovery of the vector boson. On the other hand, they now had the time to install beam extraction and the magnetic horn, and the neutrino flux was certain to be many times larger than that of the Brookhaven experiment. Also they now had a large spark chamber set-up. With great anticipation the experiment was started in June 1963. Spirits were high. It was decided to put also a bubble chamber in the neutrino beam, since it seemed that the neutrino flux was sufficiently large to produce events even in this relatively low mass device.

It was actually possible to follow the experiment very closely. The spark chambers produced pictures that were developed very quickly. Moreover, one could go into the space where the spark chambers were set up, and then see them fire if there was an event. Here a short digression about the triggering of the spark chambers.

One knew by that time that neutrino events almost always have a muon in the final state. In particular vector boson events would have a muon together with the vector boson. Everybody was very nervous about background and stray particles (coming some-how around the shielding), and it was decided to trigger the spark chamber only if a charged particle was coming out of the produc-tion region. In addition, there was an anti-trigger against the case that a charged particle entered in the beginning of the production region. Neutrino events, all things considered, are very rare events and almost anything else is overwhelmingly more frequent. Much depended on the quality of the shielding, not only in front of the detectors, but also on the side, above and below. In retrospect the shielding was very good, but that was not known beforehand. The bubble chamber cannot be triggered, it has to be fired every time the proton synchrotron discharges its protons, so if the background had been that bad that instrument would have been flooded. But such was not the case, and the bubble chamber produced about 240 reasonably clean events out of 461 000 pictures.

So, one could sit near the spark chambers and see the events coming in. This was extremely exciting. You could try to guess if an event was of the vector-boson type for example. It was like entering a new world. To me this was one of the most fascinating periods of my scientific life. It tied me forever to this profession. There is a quality to such a trip in the unknown that I can only compare to the landing of the Viking spacecraft on Mars. Those having followed that event on television may understand what I mean.

As the reader knows by now no vector boson was found in the CERN neutrino experiment. Even after a few days it became clear that if there was anything like that it was not going to be easy. This had a disastrous effect on the morale of the group. The interest in the experiment collapsed almost immediately. It looked like nothing would come of it. This was perhaps most clear in the systematic scanning of the spark chamber pictures as they came in. In the beginning everybody was hanging around the scanning tables (the pictures were projected onto some large surface where they could be looked at and measured). But after a little while almost nobody bothered to look. I believe that Bernardini and I are the only ones that have actually seen all the pictures coming from the spark chambers in the production region. We became great friends, he, the relatively old and experienced Italian experimenter, and me, the would-be theorist. I think of that time with the greatest fondness for Bernardini. And here you have the reason why I came to represent the CERN experiment at the Brookhaven conference. Bernardini, disappointed because of the negative result of the experiment, did not like to go there and instead he made me his representative.

There were many incidents in that period, and I am not going to detail any of them. In my opinion, if the spark chamber group had been more realistic, if the various participants had been more tenacious, quite interesting results would have been obtained from the data. But they had lost interest. In the end the events were not measured, and no systematic analysis was made. It is my feeling that this is an important part of scientific discovery: do not give up till the last stone has been turned. Moreover, always try to do the extra bit, go the extra distance. Ten years later we saw another example of this phenomenon, the discovery of the J/Ψ particle at

SLAC (Stanford) and Brookhaven. This particle has a mass of 3096 MeV, and could be discovered in an electron-positron collider of that much energy. This was just above the energy of such a collider at Frascati, called ADONE, whose design energy was 3000 MeV. When the people at Frascati heard about the SLAC discovery they needed only a few days to screw up the energy of the machine and observe that particle. As someone told me: they were on strike or something, but when the news broke everybody, from the cleaning personnel upwards, showed up at work on the very next day. Now why had nobody tried to do the extra thing before? I am sure some people there are still gnashing their teeth!

At the Brookhaven conference I reported that no vector boson had been seen. This was in the form of a limit. To quote verbatim: "Neglecting uncertainties in the branching ratio in the decay of the W, we conclude that $M_W > 1.3$ GeV." In other words, if there is a W in the experiment, for which there is no evidence, its mass must be larger than 1.3 GeV or else it would have been seen. The misery of the CERN experiment did not end there. Shortly after the Brookhaven conference (9–11 September 1963) another conference was organized in Siena, Italy (30 September– 5 October 1963). Also there the conclusions were presented by a theorist. I was not present as I had gone straight from Brookhaven to SLAC, but J. S. Bell, functioning as neutrino theorist (together with the Norwegian theorist J. Løvseth) did present that talk. As he felt that I had contributed to the subject he added my name to the paper, without ever showing it to me. He thought he was doing me a favour, but that was not the case! Communication between Europe and the US was in those days less easy then it is today. Anyway, physicists from all over the world were in attendance at the conference and demanded a clear statement from the CERN group: was there or was there not a vector boson in their experiment? I have heard of nightly gatherings on the top of a tower in Siena where the CERN people were tormenting each other over this question. Would they miss an important discovery? In some halfhearted way they admitted to the existence of a W.

John S. Bell (1928–1990, right) and I at CERN in Bell's office 10 years after the neutrino experiment. We were the quasi-official theorists of that experiment. We did not do very well, all things considered, because of inexperience and ignorance. After the experiment, in 1963, we both went to SLAC, where I wrote my computer program Schoonschip and he developed his famous inequalities. We also discussed other things, even wrote a paper together that was never published. He considered his work on the fundaments of quantum mechanics as a hobby, mainly to be done in the evening, at home. He told me that he intended to do away definitely with this nonsense of hidden variables, and so he did. Later he drifted more and more into this subject, and as I consider it as some sort of foolishness not good for anything having to do with the real world, I once asked him: "Why are you doing this? Does it make the slightest difference in the calculations such as I am doing?" To which he answered: "You are right, but are you not interested and curious about the interpretation?" He was right too, up to a point. While his work became very important, as it could be verified by experiment, often in this branch of physics the discussions are on the level of finding out how many angels can dance on the point of a needle. But even so: there are interesting things there.

In Ann Arbor a happening was organized on the occasion of my sixtieth birthday, in 1991. They asked Bell to talk there, but he died suddenly. When I came to CERN some time later I sat in his office and accidentally touched his computer keyboard. The screen lighted up and there was his last e-mail, to Ann Arbor: "O.K., I will sing." It was a sad moment.

I think that Bell just presented what the experimenters told him, and there is in "our" paper the ominous statement "We would be very surprised if it (the W-mass) rose as far as, say, 2 GeV." In other words, the W was there and it had a mass below 2 GeV. This was the low point in the CERN neutrino experiment.

7.6 Missed Opportunities

Opportunities were missed due to a number of factors, but I would say that the major one was the failure to analyze the spark chamber data in a systematic way. Some 2000 neutrino events had been registered in the production region in the period June–August 1963, and they have never been digested in any serious manner.

The bubble chamber group was much more serious with its analysis. They were however dealing with a much smaller number of events, about 240 in the period mentioned. It is hard to do much with this small sample, though not impossible.

What physics was there in those data? There are essentially two issues that may be discussed here, namely neutral currents and scaling. Let us start with neutral currents.

As stated before, if there are two neutrinos there is a new quantum number, namely muon number. Assuming now that the neutrinos in the neutrino beam are all muon-neutrinos it follows that in every event induced by such a neutrino there must be in the final state either a muon or a muon-neutrino. The latter case means that one has a reaction with a neutrino coming in and one going out, and no muon. This type of event is called a neutral current event. While one cannot see the neutrinos one can see the other products of the collision, such as the nucleus breaking up in addition to new particles such as pions. The difficulty is that this is not a very clear signature: no long muon track, no electron shower. It looks in the first instance quite a lot like an event that would be induced by a stray neutron, hitting a nucleus and making it come apart. Thus identifying this type of event requires a clear background analysis. To see how such events look like take all the events containing a muon and take away the muon. There were 2000 examples of events with a muon.

Theoretically there was a rather heavy bias against this type of events (without a muon). In quite different circumstances things like that, involving two neutrinos and no electron or muon had been found to be absent with a high degree of certainty. By 'involving two neutrinos' we mean events where there were two neutrinos, either one neutrino coming in and one going out, as in the neutrino experiment, or two going out, as in certain decay type reactions. As we will see, those things are theoretically very similar. Since no one had any idea about the details of neutrino interactions no one thought twice about this type of reaction. Whether they were there or not was not a burning issue. Lee and Yang had not made the point with any force. Neutral current events are in fact mentioned in their article, but not more than that. At the end of their influential article they literally state "the question of a neutral vector boson will not be examined here". Consider the expression 'neutral vector boson' as a synonym for 'neutral currents'. Here I can only repeat: there was not much interest in that question at that time, something that changed after 1971.

Experimentally it had been made sure that even if there were such events they would not be discovered. Living in fear of the background there was the muon trigger, requiring a muon leaving the production region. If there was a neutral current event the spark chamber would not fire. At some meeting of the group the issue of running without a muon trigger was raised, but it was voted down. Even then there was something that could possibly be seen in the data, as there were sometimes two events on one picture. In any case, no neutral current events were ever reported.

The bubble chamber was in a different position. Here there was no muon trigger. In fact, we know now that there must have been a substantial number of neutral current events in the bubble chamber pictures. An analysis was made, but (1) no one was

Gilberto Bernardini (1906–1995). In 1963, when I started taking an interest in the neutrino experiment, Bernardini was essentially the boss. I got in because I had redone calculations concerning *W*-production (first done by Lee, Markstein and Yang) adapted to the CERN experiment requirements. Bernardini and I became friends instantly, although one can of course not ever become as good friends with an Italian as another Italian. One lacks the refinement in language, choice of words, and knowledge of Italian literature. Bernardini was a very cultured man.

From time to time we used to walk the CERN corridors. He would occasionally put his arm around my middle, which embarrassed me greatly. It must have been a remarkable sight, he the little Italian (from my perspective) and me, the much younger rather blocky Dutchman. We shared however one thing: passion for physics. When the initial excitement was over most spark chamber people did not show up in the evening or at night. However Bernardini and I were there every night, looking at the spark chamber pictures and hoping to see the *W*, or even better, the unexpected. It was an exciting period.

The rather famous picture of Bernardini above shows one of those things. The neutral current ratio R was given as to be less than 5% (the equation below his arm at the level of his middle). In actual fact the number is about 15%. This error was due to some misidentification in the bubble chamber pictures. A correcting article was published before gauge theories, demanding neutral currents, became popular (after the 1971 Amsterdam conference).

interested and (2) errors were made. Today some (including me) believe that neutral current events could have been established on the basis of those bubble chamber data, but just marginally.

After 1971 neutral current events became overwhelmingly important as they would testify to the gauge structure of weak interactions (this will be discussed later). Let me digress on this for a moment.

Neutral current events are like neutrino events with the muon missing. Background events induced by stray neutrons are different in a number of ways. Firstly, neutron induced events look different, the secondary products tend to be much more spherically distributed. Second, since neutrons are absorbed rather easily, these neutron events tend to be located in the first part of the bubble chamber. If the bubble chamber is too small, that effect may not be sufficiently manifest. After 1971 when the very large French heavy liquid bubble chamber (called Gargamelle) was placed in the neutrino beam this analysis became feasible and indeed neutral current events due to neutrinos were established.

The issue of scaling is more complicated. This has to do with the probability of a neutrino event depending on the energy of the neutrino, and the question of elastic versus quasi-elastic events.

A neutrino basically collides with a neutron in a nucleus, and whether or not the remainder of the nucleus remains intact is of little consequence. Of interest is the basic mechanism, the neutrino colliding with a neutron. The following type of event is called a quasi-elastic event:

$$\text{neutrino} + \text{neutron} \rightarrow \text{muon} + \text{proton}$$

Normally an elastic event has the same particles in the final state as in the initial state. In some global way that is what we have here if we see the muon and its neutrino as two members of one and the same family, and similarly for the neutron and the proton. An inelastic event is when extra particles are created, for example

$$\text{neutrino} + \text{neutron} \rightarrow \text{muon} + \text{proton} + \text{pion(s)}$$

For higher neutrino energy inelastic events become much more probable then quasi-elastic events. This is well understood today because we know that the neutron is made of quarks, and the reaction is basically a neutrino-quark collision. A sufficiently high-energy neutrino will simply break up the neutron, which will show up as debris in the form of pions as seen in inelastic collisions. Just as we ignored what the other nucleons in a nucleus do we should ignore what the other quarks in the neutron do. That is how we see it now. The break-up of a neutron is more complicated than that of a nucleus, but there is no basic difference. But this was not understood in the old days, and in 1963 no one had an inkling about quarks.

The figure below shows a plot of the number of neutrino events seen depending on the energy of the neutrinos. It is assumed here that the neutrino flux is the same for neutrinos of all energies (which of course is not true, but that can be taken into account). For low energy neutrinos only the quasi-elastic reaction is possible. Initially the number of such events goes up strongly as the energy of the neutrinos increases,[a] but then levels off to a constant. However, for higher neutrino energy the inelastic reaction becomes possible, and as the energy goes up reactions with more and more secondary pions are seen. All in all, the inelastic events tend to compensate for the leveling off of the quasi-elastic reaction, such that the strong increase at low energy persists to higher energies if inelastic reactions are included. This whole behaviour can be understood by assuming that the neutrinos essentially collide with a quark inside the neutron or proton, and at higher energies the neutron or proton breaks up, which is manifested by extra pions being emitted. The basic reaction, however, always remains the same (neutrino-quark scattering), and one knows of that reaction that its probability increases with the neutrino energy in the manner seen.

---

[a]Proportionally to the neutrino energy squared.

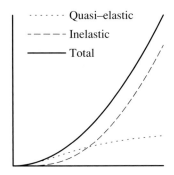

In the figure we have cheated a bit to make the point. At low energies only the above mentioned quasi-elastic event involving initially a neutron can happen. A neutrino impinging on a proton cannot produce a quasi-elastic reaction because of charge conservation. The neutrino becomes a negatively charged muon (it must because of muon number conservation) and therefore the positively charged proton would have to become a doubly charged nucleon, which does not exist. The only nucleons with a mass below 1000 MeV are the proton and the neutron. For inelastic events there is no such problem, there may simply be an extra π^+. Thus inelastic events also happen with initially a proton, and that must be taken into account when making the plot.

7.7 Epilogue

The spark chamber pictures of the production region remained at CERN in some cupboard, and no one worked on them. In 1971 I visited CERN and decided to have a look at them, because it had become clear to me that neutral currents were important in connection with the developing theory of weak interactions. I looked at the place where I had last seen them, but I could not find anything. Eventually I discovered that they had been burned. To this day I find that one of the most incredible things about that experiment.

$$\widehat{8}$$

The Particle Zoo

8.1 Introduction

Around 1960 the situation in particle physics was very confusing. Elementary particles[a] such as the photon, electron, muon and neutrino were known, but in addition many more particles were being discovered and almost any experiment added more to the list. The main property that these new particles had in common was that they were strongly interacting, meaning that they would interact strongly with protons and neutrons. In this they were different from photons, electrons, muons and neutrinos. A muon may actually traverse a nucleus without disturbing it, and a neutrino, being electrically neutral, may go through huge amounts of matter without any interaction. In other words, in some vague way these new particles seemed to belong to the same group of particles as the proton and neutron. In those days proton and neutron were mysterious as well, they seemed to be complicated compound states. At some point a classification scheme for all these particles including proton and neutron was introduced, and once that was done the situation clarified considerably. In that era theoretical particle physics was dominated by Gell-Mann, who contributed enormously to that process of systematization and clarification. The result of this massive amount of experimental and theoretical work was the introduction of quarks, and the understanding that all those 'new' particles as well as the proton

---

[a]We call a particle elementary if we do not know of a further substructure.

Luis Alvarez (1911–1988). After Glaser came up with the idea of a bubble chamber Alvarez was quick to realize the potentialities of such an instrument. With considerable energy he put himself to the task of building bubble chambers, and to use them for physics purposes. With his group of very talented engineers and physicists (the distinction was not always clear) at Berkeley he started constructing a then relatively large hydrogen bubble chamber (10 inch = 25 cm long), with which a large amount of physics was done. They discovered many of the particles mentioned in this section. Alvarez received the 1968 physics Nobel prize.

In a subsequent daring step the Berkeley group went on to construct a much larger hydrogen bubble chamber (72 × 20 × 15 inch = 183 × 51 × 84 cm) for the then large sum of $2.5 million. The problems were huge: liquid hydrogen (or deuterium) had to be kept at a temperature of −250°C, and the magnet surrounding the bubble chamber was very large (100 tons, using some 2 Megawatts to power it).

The first very significant result obtained with the 72-inch chamber was due to Pevsner and his group at Johns Hopkins University. The chamber (filled with deuterium) was exposed to a beam of pions from the Bevatron (a 6-GeV accelerator in Berkeley) and photographs were taken and sent to Johns Hopkins. The result was the discovery of the η, which particle completed the octet of mesons as described in this Chapter.

The relation of Alvarez with the then director of LBL (Lawrence Berkeley National Laboratory), Edwin MacMillan, deteriorated to the point that it interfered with the physics done. So it goes.

Scanning table at CERN in 1972. These devices were used at all institutions engaging in particle research. Rolls of film would be recorded during some run at one of the big accelerator laboratories and then scanned and analyzed at the various university laboratories. Up to a million of such pictures were recorded, and one can see the huge and rather dull work associated with that. The physicist became more of a manager rather than an experimenter. The scanning was usually done by girls who often did not know anything about the subject.

This kind of physics, while a necessity for progress, tended to make particle physics dull and uninteresting. At the scanning table the data was recorded on magnetic tapes for further processing by computers, and things became interesting again after computers processed the data and summarized the results in graphs and histograms. Then patterns could be found and new particles discovered. The new particles, all of them highly unstable, would decay in a very short time, and they were established through analysis of the decay products. For example, Pevsner and his group at Johns Hopkins University obtained films from the 72 inch hydrogen bubble chamber at Berkeley exposed to a pion beam. The pions colliding with protons in the bubble chamber gave rise to events with many particles coming out. Pevsner and co. then searched for combinations involving three pions, and tried to figure out if the three pion configurations were consistent with the decay of a single particle (the η). The curvature of the tracks (due to a magnetic field in the bubble chamber) allowed the determination of the particle energies, and from them the mass of the η (about 550 MeV). Not all three-pion systems are due to η decay, so this was actually a lot harder than it seems.

and neutron were various bound states of quarks. So this is what this Chapter is about: bound states of quarks. There are many of them, and they form what we may call the particle zoo. They are particles, but not elementary particles. Some of them have been mentioned before, namely pions and kaons.

It must be well understood that although hypothetical particles called quarks could theoretically be used to understand all these states as bound states of these quarks, there was nonetheless at that stage no evidence that the quarks were actually real particles, with a well-defined mass. That changed completely after 1967, when experiments at SLAC showed that inside protons and neutrons there were point-like things. This will be discussed in Chapter 11.

8.2 Bound States

Thus at this point the big complication was that for some reason, even now not yet completely understood, the quarks cannot occur by themselves, free. They occur only in bound states. That was difficulty number one. Furthermore, the way that the quarks are bound differs quite a lot of what is seen in other known bound states such as atoms and nuclei, and it took quite some time before this was understood. That was difficulty number two, which we shall describe now.

In a hydrogen atom the constituents (one proton and one electron) are still easily recognized. The binding energy is relatively low, so that the total energy of the atom is very close to the sum of the energies contained in the masses of the electron and the proton.

To be precise, the masses of the electron and the proton are about 0.511 MeV and 938.272 MeV respectively, and the binding energy is -13.6 eV $= -0.0000136$ MeV. The binding energy is negative, you must add energy to tear the atom apart. Clearly the binding energy is next to nothing compared to the mass energies, and the mass of the hydrogen atom is in good approximation equal to the sum of the electron and proton masses.

For nuclei the story is quite similar, except that the binding energy is much larger. However, it is still small compared to the masses of the protons and the neutrons in the nucleus.

For helium, for example, the nucleus contains 2 protons and 2 neutrons, and using the mass values 938.272 and 939.563 MeV gives 3755.67 MeV for the mass energy of the helium nucleus. For the helium nucleus the binding energy (equal to minus the energy needed to tear that nucleus apart into its constituent protons and neutrons) is − 28 MeV. Thus the binding energy is about 0.7% of the total energy. The nuclear binding energy is slightly different from nucleus to nucleus, and is usually quoted in terms of binding energy per nucleon. For helium that is − 28/4 ≈ − 7 MeV.

Thus it is quite easy to count how many protons plus neutrons there are in a given nucleus, simply by measuring its mass. That makes it easy to realize that nuclei are bound states of protons and neutrons. But with bound quark states that is a very different matter.

Bound states of quarks are complicated structures. The reason is that the gluons, responsible for the strong interactions between the quarks, also interact with themselves, and there are big globs of gluons that keep the quarks bound. While the gluons themselves are massless, they do have energy, and the gluon globs are quite energetic and thus contribute to the mass of the bound state. The quarks are embedded in gluons. The masses of the quarks are only a small part of the mass of the bound states. For example, the proton has two up quarks and one down quark, which accounts for about 15 MeV of the mass of the proton, 938.27 MeV. Thus the gluon blob contains some 923 MeV! It is very hard to even speak of binding energy in those circumstances. Moreover, it is not possible to separate the proton into its quark constituents. As the quarks are moved apart more and more gluon matter builds up between the quarks, requiring energy, and that energy keeps on increasing no matter how far the quarks are separated. This kind of binding mechanism is totally unknown elsewhere, and that made it so hard to recognize the real state of affairs.

It is obviously not easy to determine the quark masses in these circumstances. A certain amount of not too clear theory goes into that, and consequently there are quite large uncertainties here, in particular for the up and down quark. However, information on the mass difference between the up and down quark mass can be guessed from the mass difference between proton (*uud*) and neutron (*udd*); 1.291 MeV. Proton and neutron are very similar in their quark-gluon structure, and the main difference is in electric charge. The energy related to the electric force must be taken into account, and the up-down quark mass difference is estimated to be somewhere between 1.5 and 4 MeV.

Matters change when heavier quarks are involved. Bound states containing heavy quarks were discovered after 1967, so these states did not play any role in the question of hypothetical versus real quarks. The heavy quarks are the charmed, bottom and top quark, with masses of approximately 1.3, 4.5 and 175 GeV (1 GeV = 1000 MeV). These masses are quite large compared to the energy contained in the gluon blobs, and it is easy to guess how many of these heavy quarks are contained in any bound state. In 1974 the first bound state involving heavy quarks was discovered and identified as a new particle simultaneously at SLAC (Stanford) and BNL (Long Island). The people at SLAC called it a ψ, those at Brookhaven a J, and till today we are saddled with this dual name. This J/ψ particle, with a mass of about 3000 MeV, was later established to be a bound state of a charmed quark and an charmed quark.[b] The mass of the J/ψ is 3096 MeV, as compared to the sum of the quark masses of about 2600 MeV. Apparently there is here about 500 MeV in the gluon

---

[b]Reminder: the bar indicates the antiparticle.

blob. The amount of energy in the glue of the quark bound states varies from case to case, and is generally in the range of 120 to 1000 MeV.

The foregoing makes clear that it was quite difficult to recognize the observed particles (those involving up, down, and strange quarks only) as bound states of varying numbers of quarks and antiquarks.

8.3 The Structure of Quark Bound States

Today a proton is understood as a glob of gluons with three quarks swimming in it. One might ask if such gluon blobs could also exist without any quarks in them, and in fact that has been suggested. Extended experimental searches have not produced convincing evidence for such particles, tentatively called glue-balls. Somehow the quarks seem to be a necessary ingredient.

The branch of physics that is about gluons and their interactions with themselves and with quarks is called quantum chromo-dynamics (QCD). It is a very complicated subject, and it will not be discussed in any serious way in this book. The complications arise because, as mentioned above, the various types of gluons interact with each other in a complicated way. It is due to this that one can have large blobs of gluons that seem to resemble wads of chewing gum. The analogy goes even further: when considering a two-quark bound state one may try to take it apart. What happens is that, when separating the quarks, a string of glue appears to form between the two quarks. As if trying to tear a piece of chewing gum apart. The difference is that the chewing gum will break at some point, while the gluon glob just keeps on stretching. The peculiar thing about it is that the force with which the two quarks are held together apparently remains roughly the same, no matter how much they have been pulled apart. That at least is more or less what most particle physicists think today, although the evidence for this precise constant behaviour is not very substantial. In any case, one can apparently never get the two quarks separated. A lot of experimentation and theory has gone into that, but these gluon strings remain difficult objects. They are an approximate description of a complex situation. People have idealized and abstracted these strings of glue to string-like objects that have no quarks and are not glue either, and that has given rise to string theory, studied widely. However, there is no evidence of any kind that Nature uses strings other than in the approximate sense of gluon matter between quarks relatively far apart.

Let us now describe the above in more detail. In the past, when quantum mechanics was introduced, the first important system to which the theory was applied was the hydrogen atom. One started from the known electric attractive force between electron and proton. This force, the Coulomb force, is known to fall sharply as the distance is increased, and to be precise it falls quadratically with that distance. So if the electron and proton are at some distance there will be an attractive force, and then moving the electron out to twice that distance the force becomes four times smaller.

Now, using this Coulomb force law, quantum mechanics predicts the various bound states for an electron and a proton, and these bound states are the excited states of hydrogen. They correspond to the electron circling in higher orbits. If now an electron circulating in a higher orbit around the proton drops to a lower orbit it will emit a photon. The energy of that photon is precisely equal to the difference in the energy of those two bound states. Thus by observing the energy of the photons emitted by hydrogen after being put in an excited state (this can be done by bombarding the hydrogen atom with electrons) one may precisely establish the energies of the excited states, that is the bound states with the electrons in higher orbits. The experimentally observed optical spectrum of hydrogen agreed very well with the energies of the bound states found when using the Coulomb force law in the quantum mechanical calculations. Conversely, if one had not known about the Coulomb force law, one could have deduced that law by observing the spectrum and then trying to find which force law would reproduce the observed spectrum of photon energies. That is the procedure which one tried to apply in connection with the quark bound states.

There is some (scanty) evidence that associated with a given bound state of quarks there were higher mass bound states, with higher spins. Making a plot of these bound states, plotting spin versus the square of the mass, something like a straight line seemed to appear.

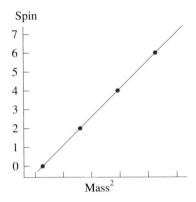

Such a line is called a Regge trajectory (after the Italian physicist T. Regge). Using the procedure sketched above such a spectrum of bound states of two quarks can be understood as due to a force that would be independent of the distance between these quarks. That force was interpreted as due to a string-like configuration of quarks and gluons.

In time the Regge trajectories thus became the cradle of string theory. Nowadays the Regge trajectories have largely disappeared, not in the least because these higher spin bound states are hard to find experimentally. At the peak of the Regge fashion (around 1970) theoretical physicists produced many papers[c] containing families of Regge trajectories, with the various (hypothetically straight) lines often based on one or two points only!

---

[c]Just like these days on the subject of strings.

Yoichiro Nambu (1921). Nambu interpreted the success of the Regge idea in terms of a force between quarks. He also had a large influence on the development of quantum chromodynamics; together with a collaborator, Han, he essentially introduced quark color charges. Not only that, they then also introduced what we now call gluons. Their work was yet a far cry from the rather elegant theory of quark and gluon interactions (quantum chromodynamics) that is today contained in the Standard Model, but the basis for a considerable part of the theory was undoubtedly in their paper.

Another important contribution by Nambu (together with Jona-Lasinio) is the idea of a neutral field in the vacuum. While such a field would not be observable by direct experimentation, it could explain a number of observed facts. This idea became the basis of the work of Brout, Englert and Higgs (see Chapter 10) that was of fundamental importance in connection with gauge theories.

Somewhere in the nineties I had an unexpected encounter with Nambu. I had developed some equation that contained a relationship between the top quark mass and the Higgs particle mass, both particles then still to be discovered. If the top is sufficiently heavy that relation becomes very simple: the Higgs is twice as heavy as the top. At that point, at Fermilab, I ran into Nambu who not only had arrived at the same equation, but in addition came up with the idea that the Higgs might thus be a bound state of a top and an antitop quark (which indeed would put the Higgs mass at about twice the top mass). We went together to question the experimenters about the state of affairs, but then, as now, there was no answer. We are still waiting for the Higgs.

It is for our purposes quite pointless to describe the multitude of bound states observed. The discussion will be restricted to bound states of the light quarks, that is the up, down and strange quarks, and even more narrowly to some subset of these bound states, namely the states of lowest mass. Those states were experimentally discovered in the period 1948–1965. Mainly quark-antiquark bound states, called mesons, and three-quark bound states, called baryons will be reviewed. Bound states containing heavy quarks (charm, bottom and top) will be discussed briefly after that.

8.4 Spin of a Bound State

A bound state is just another particle, just as an atom may be considered a particle. Any particle has a spin that may be considered as an internal state of rotation. It is really like a spinning tennis ball. However, on the particle level there are quantum effects, meaning here that only certain amounts of rotation, of spin, are possible. All spins must be integer or half-integer multiples of a certain basic quantity. That basic quantity will be taken as the unit, so spins can take the values 0, $\frac{1}{2}$, 1, $\frac{3}{2}$, 2, $\frac{5}{2}$, etc. The spin of a bound state is equal to or between the sum and difference of the spins[d] of its constituents plus an integer amount. The extra integer amount can be seen as a rotation of the constituents around each other. Negative spin does not occur, to us spin is simply the amount of rotation, and that can be zero but not less than zero. So this is the picture: the total amount of rotation is the internal rotation of the quarks themselves (the spin of the quarks) plus the spin due to these quarks rotating around each other. It is a simplified picture, because the gluon matter may (and does) rotate as well, but altogether one obtains the result described, as if ignoring the gluon glob.

---

[d]However always integer or half-integer if the sum is integer or half-integer.

8.5 Mesons

Mesons are defined as bound states of one quark and one anti-quark. Both quark and antiquark have spin $\frac{1}{2}$. The spin of a meson can be 0, 1, 2, etc. We start with low mass spin zero particles.

Considering only bound states of up, down and strange quarks there are nine possibilities. These possibilities are listed below. The first line lists the quark antiquark combinations, the second line the symbols of the experimentally found particles that appear to correspond to these combinations. As usual, the bar indicates an antiquark, thus for example $\bar{u}$ is the antiup quark or up quark.

$d\bar{s}$	$u\bar{s}$	$d\bar{u}$	$u\bar{d}$	$s\bar{u}$	$s\bar{d}$	$d\bar{d}$	$u\bar{u}$	$s\bar{s}$
K^0	K^+	π^-	π^+	K^-	$\overline{K}^0$	π^0	η	η'

The color charge of the quarks (see Chapter 2) plays no role in this discussion; the bound states are color neutral. This means that if there is for example a red quark, there is also an anti(red quark). The bound state will be a mixture of the possible color combinations red–antired, green–antigreen and blue–antiblue.

The pions (π) and kaons (K) have been mentioned before, in Chapter 6. These particles were copiously produced at the first big machines (CERN, BNL), and became the subject of intense experimentation. All particles shown on the second line were discovered before it was realized that they were bound states of a quark and an antiquark, and the names shown are those given in the pre-quark era. The electric charges of these particles are as shown, if not indicated (η and η') they are zero.

The table is strictly speaking not correct, because the π^0, η and η' are not precisely the bound states listed above them, but certain mixtures. For example, the π^0 is a mixture of $d\bar{d}$ and $u\bar{u}$. There is no need to worry about that here.

In 1961 all these particles were classified in a particular manner, best shown in a figure. This most remarkable figure, introduced by Gell-Mann in his paper entitled "The eightfold way", immediately took hold in particle physics. As we will see it is suggestive of a construction built up from triangles, and that is indeed what led to the introduction of quarks in 1964. The nine particles are grouped into an octet (8 particles) and a singlet.

particle	mass (MeV)	lifetime (seconds)
π^+, π^-	140	2.6×10^{-8}
π^0	135	8.4×10^{-17}
K^+, K^-	494	1.2×10^{-8}
K_S	498	0.89×10^{-10}
K_L	498	5.2×10^{-8}
η	548	5.6×10^{-19}
η'	958	$3.3 \times 10^{2-21}$

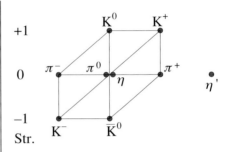

Spin 0 Meson Octet and Singlet

In this figure the particles are arranged by strangeness and charge; for our purposes the strangeness of a particle is determined by the number of strange quarks in that particle. For every strange quark count -1, and $+1$ for its antiparticle, the strange quark. For example, K^- has one s quark, and thus has strangeness -1. The strangeness is the same for particles on the same horizontal line; charge is the same for particles on the same vertical line. The classification into octet and singlet is related to the behaviour of the bound states under exchange of the quarks. The η' is supposedly an equal mixture[e] of $u\bar{u}$, $d\bar{d}$ and $s\bar{s}$. It remains the same thing if the quarks are interchanged, for example, if the d and $\bar{d}$ are interchanged with s and $\bar{s}$. Particles in the octet interchange with each other, for example that same quark interchange ($d \leftrightarrow s$ and $\bar{d} \leftrightarrow \bar{s}$) exchanges K^0 and $\overline{K}^0$.

---

[e]There is some further mixing, but that is of no relevance here.

Customarily one draws this figure in a slightly more symmetrical way. The charges of particles on the same diagonal line (upper left to lower right) are then the same.

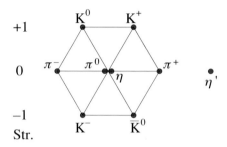

Spin 0 Meson Octet and Singlet

If the quarks had all the same mass, all particles in the octet would presumably have the same mass. However, the mass of the strange quark (the strange quark has the same mass) is higher than those of the up and down quarks, and thus the kaons, containing one strange quark or antiquark are heavier than the pions. The η and η' also contain a strange quark and an anti(strange quark) and are even heavier. As indicated in the table all of these particles are unstable.

K_S and K_L are certain mixtures of K^0 and $\overline{K}^0$. If you are confused by all this mixing business you are in good company: it took quite some time before all this was unraveled and understood. Now we know, but it is never really easy. Luckily there is rarely any need to go into details, at least not within the framework of this book.

The charged pions and kaons decay with relatively long lifetimes (of the order of a few one-hundredths of a micro-second), such that they actually make tracks that can be observed and measured. The neutral pion decays very fast, but by very refined methods it has nonetheless been possible to establish a path over a small distance prior to decay. The distance covered is of the order of a micron (one micron is a millionth of a meter). The neutral pion decays almost always into two photons.

The η, discovered by Pevsner and his group in bubble chamber data around 1960, is very unstable, and decays so fast after production that no track can be seen in the usual detection instruments. Such particles are established purely on the basis of the mass-shell relation as described before. The η decays mainly into two photons or three pions, and by carefully measuring the momentum and energy of the pions (or the photons) one establishes the mass of the η from the total energy and momentum of the decay products. The particle is established by the fact that in many events the same mass value results.

The above mentioned states are bound states where the spins of the quark and the antiquark point in opposite directions. Also there is no motion of the quarks around each other, which makes for relatively simple bound states. Almost as simple are the bound states without relative motion, but where the quark spins point in the same direction. Then the total spin is 1. Here is the corresponding set of spin 1 particles as observed.

$d\bar{s}$	$u\bar{s}$	$d\bar{u}$	$u\bar{d}$	$s\bar{u}$	$s\bar{d}$	$d\bar{d}$	$u\bar{u}$	$s\bar{s}$
K^{*0}	K^{*+}	ρ^-	ρ^+	K^{*-}	$\overline{K}^{*0}$	ρ^0	ω	ϕ

The spin 1 particles may also be arranged into an octet and a singlet. Here the lifetimes are not given, as they are very short.

particle	mass (MeV)
ρ^+, ρ^-	770
ρ^0	770
K^{*+}, K^{*-}	892
K^{*0}, $\overline{K}^{*0}$	892
ω	782
ϕ	1020

Spin 1 Meson Octet and Singlet

8.6 Baryons

The particles to be described here are called baryons, and they are bound states of three quarks. The situation with respect to spin is more complicated than in the meson case, and will not be described in any depth. The best known particles can be separated into two groups, containing respectively eight particles of spin $\frac{1}{2}$ (two quarks with spin up and one with spin down) and ten particles of spin $\frac{3}{2}$ (spins of all three quarks in the same direction). The group of eight particles fits nicely into an octet like in the meson case, the group of ten (decuplet) fits into a new type of figure. There are no singlets.

In the case of mesons the antiparticles are in the same octet as the particles. Thus K^- and K^+ are each other's antiparticle, but they are in the same octet.

The baryons are bound states of three quarks, for example the proton has two up and one down quark. The antibaryons contain antiquarks, thus the antiproton contains two anti(up quarks) and one anti(down quark). The antibaryons thus form an octet and a decuplet by themselves. With the rule that particles of less strangeness appear lower in the figures it follows that in the case of the antibaryons the figures must be drawn upside down. This because the particles containing an anti(strange quark) have strangeness $+1$ and must be placed above the other particles that have zero or negative strangeness. Thus the Ω^-, three strange quarks, has strangeness -3 and charge -1, while the $\overline{\Omega^-}$, three anti(strange quarks), has strangeness $+3$ and electric charge $+1$. The antibaryons have the same mass and lifetime as the baryons. But let us now return to the baryon octet and show the list and the corresponding figure.

The proton is of course stable, or you would not be reading this. The neutron lives very long, about 10 minutes, due to the fact that the energy difference between proton and neutron is quite small (about 1.3 MeV). A little binding energy in a nucleus goes a long way to compensate this and that makes the bound neutrons stable. Most nuclei up to uranium, containing many

particle	mass (MeV)	lifetime (seconds)
P(roton)	938.3	stable
N(eutron)	939.6	887
Σ^-	1197.4	1.5×10^{-10}
Σ^0	1192.6	7.4×10^{-20}
Σ^+	1189.4	0.8×10^{-10}
Ξ^-	1321	1.6×10^{-10}
Ξ^0	1315	2.9×10^{-10}
Λ	1115.7	2.63×10^{-10}

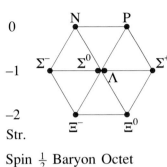

Spin $\frac{1}{2}$ Baryon Octet

neutrons, are stable. The other particles, with the exception of the Σ^0, live long enough to traverse a measurable distance in the usual detection instruments. The Λ is neutral, and in a bubble chamber it can be observed when it decays into a proton and a negative pion. That gives a 'V', some distance away from the point where the Λ was produced. In the early days (fifties) the Λ was called a V-particle. It contains one strange quark.

Now the spin $\frac{3}{2}$ baryon decuplet. The very short lifetimes are not indicated. Going down amounts to replacing a down quark by

particle	mass (MeV)
Δ	1232
Σ^*	1383
Ξ^*	1532
Ω^-	1672.5
Mass differences:	
$\Sigma^* - \Delta$	151
$\Xi^* - \Sigma^*$	149
$\Omega^- - \Xi^*$	140

Spin $\frac{3}{2}$ Baryon Decuplet

a strange quark. Thus the Σ contains one strange quark, the Ξ two and the Ω three. Correspondingly, going down, one would assume the mass to increase by something close to the strange quark mass. From the table that mass appears to be around 150 MeV. On the other hand, the mass difference between a pion and a kaon is 350 MeV, and it is clearly not easy to pinpoint the strange quark mass. It is probably somewhere between 60 and 170 MeV. Not knowing any better is a testimony to our poor understanding of the quark bound states.

Historically the Δ was discovered by Fermi, in 1952. It is the earliest highly unstable particle discovered. It took some time before physicists realized that such a highly unstable system must still be considered a particle. It is just very unstable.

8.7 Exotics

Here we will discuss a few quark bound states involving the charm, bottom and top quarks. The earliest detected is the J/ψ, a charm-charm bound state also called charmonium, with a mass of 3097 MeV. It was the first discovery of a state containing a charmed quark. Important are the B-particles, containing one bottom or bottom quark: B^+, B^- B^0 and $\overline{B}^0$, all with a mass of about 5279 MeV. These B-particles are the subject of intensive study, because their decay modes may give information on the fourth parameter of the CKM rotation (see Chapter 3). That is the parameter related to CP violation, not discussed in this book.

The first sign of a bottom quark was the discovery of the Υ (or bottonium), mass 9460 MeV. From this the mass of the bottom quark was guessed to be in the region of 4.1 to 4.5 GeV. The top showed itself in certain events observed at Fermilab around 1995. From these events a mass of about 175 GeV was deduced. The wildly varying masses of the various quarks are really baffling: 5, 10, 200, 1300, 4500, 175 000 MeV!

8.8 Discovering Quarks

The state of affairs in 1964 was as described above: particles could be grouped into multiplets as shown, and very convincingly, open spots were filled in by experiment. One of the last particles discovered was the Ω^- in the baryon decuplet; it was finally seen in a bubble chamber experiment at Brookhaven. The mass was predicted rather precisely, simply by assuming that the Ω^- mass would be another 150 MeV up from the (known) Ξ^* mass. And indeed, there it was.

At this point it was completely natural[f] to assume that all these particles are bound states of more elementary objects, and this was how quarks were invented (by Gell-Mann, and Zweig). The idea is truly simple: it is quite obvious that the multiplets shown have basic building blocks, namely triangles. The convention is as described before: strangeness decreases when going down, charge increases when going to the right. Then for antiquarks an upside down triangle must be used, as shown in the figure.

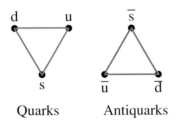

Quarks   Antiquarks

The particles in these triangles were called quarks and antiquarks, and it is quite easy to see how the nine spin zero mesons can be obtained by combining a quark and an antiquark. Start with a quark triangle (the left triangle in the figure above), and then put an antiquark triangle (an upside down triangle) onto each of the vertices such that the centers of the antiquark

---

[f]That does not mean it was easy. Intellectual courage was needed to introduce never-seen particles with a non-integer charge.

triangles are precisely on the vertices. Presto, an octet and a singlet appear as shown in the next figure.

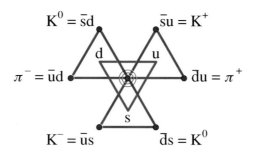

The triple circled point in the middle has the multiplicity three, as each of the three antiquark triangles has a point there. Of these three two are part of the octet and one of them is a singlet all by itself.

This procedure shows which quarks are contained in a given state. Just check which quarks have been used to generate the point. For example, the leftmost point contains the leftmost antiquark of the green triangle, which is the $\bar{u}$ quark, and the leftmost quark of the red triangle, the d-quark. If the multiplicity at some point is larger than 1 then the resulting states will usually be mixtures. The center of the figure has the multiplicity three, and the resulting particles will be mixtures of $\bar{u}u$, $\bar{d}d$ and $\bar{s}s$. The particles observed are the π^0, the η and the η', and they are thus mixtures. In the previous figure of the meson nonet we have drawn the η' on the side, but its quark content is that corresponding to the center of the picture shown here.

As shown above the spin 0 meson octet and singlet can thus be interpreted as quark–antiquark bound states, with the quark spin opposite to the antiquark spin resulting in a total spin of 0. The spin 1 meson octet and singlet must be understood as a similar construction, except that now the spins of the quark and antiquark contained in a given state point in the same direction.

The situation with the baryons is somewhat more complicated, but the figures show quite clearly how the triangle remains the basic building block. Combining the baryons as done for the mesons would give 3 × 3 × 3 = 27 states, and it is not directly clear how this reduces to an octet and a decuplet (18 particles in total). The reason is that one must make groups of particles that transform into themselves when exchanging the quarks, such as for example a nonet splitting up in an octet and a singlet. It would carry us too far to dish this out, and it is not that urgent anyway.

The following figures nonetheless give an idea. The first quark triangle is dashed black. Drawing triangles around the corner points of the black triangle one obtains the second figure with the dashed red, blue and green triangles. Now add the third quark. Take the dashed blue triangle and draw a blue triangle around each of the corners. Similarly with the dashed red and green triangles. The result is shown in the third figure. Some of the triangles have been made a little smaller, for better visibility.

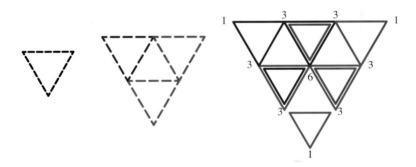

Note that the colors in these figures have nothing to do with the quark color charge, discussed in Chapter 2. Colors have been used here to make it easier to recognize the construction.

Several points in this plot are produced several times. The numbers show the multiplicity for the various points. For example, the second point in the top row (multiplicity 3) is touched by two blue and one green triangle and the point in the center (multiplicity 6) is touched by 2 red, 2 blue and 2 green triangles. The result is one decuplet, two octets (the points marked with 3 or 6) and one singlet. Indeed,

$$10 + 8 + 8 + 1 = 27 \,.$$

One of the octets is the baryon octet. The remaining octet and singlet will not be discussed.

Consider the above as a simplified discussion, as there are complications relating to the spin structure. Note that the particles of the decuplet have spin $\frac{3}{2}$, while the particles of the baryon octet have spin $\frac{1}{2}$.

In Nature one does not observe states corresponding to bound states of two quarks as would correspond to the second figure (the dashed colored triangles) in the drawing above. At the time this was not understood. It was not known then that each quark comes in three varieties coded red, blue and green. With a quark and an antiquark one can make a neutral quark color state, for example the π^- can be understood as the bound state of an anti(red-up quark) with a red-down quark. With two quarks you cannot make a state that is neutral with respect to quark colors. You can do it with three quarks: make them red, blue and green (which is white) in every bound state. Why only neutral quark color states appear in Nature is not completely understood, but we have a good idea about it. It is due to the interactions of the gluons with quarks and with themselves.

I may perhaps terminate this section with a little anecdote. When quarks were not immediately discovered after the introduction by Gell-Mann he took to calling them symbolic, saying they were indices. In the early seventies I met him at CERN and he again said something in that spirit. I then jumped up, coming down with some impact that made the floor tremble, and I asked him: "Do I look like a heap of indices?" This visibly rattled him, and indeed after that he no more advocated this vision, at least not as far as I know.

8.9 Triplets versus Doublets and Lepton-Quark Symmetry

Here is an occasion to illustrate what is easy and what is hard in physics. To extend a theory, an idea, that is in general easy. When an idea is launched for the first time you will often see it followed up by many articles, one grander than the other, and most of them, seemingly, much clearer and brilliant than the one containing the original idea. In other words, it is not always directly visible which paper was the important one. It is this odd idea, the thing orthogonal to everything else that is so hard to produce. Usually after it is introduced everyone will say: "of course". The following example is perhaps not the very best possible one, but it may illustrate the point. Around 1970 most particle physicists were thinking in terms of Regge trajectories and SU3. Now SU3 is the scheme of octets and decuplets shown above, and we now understand this multitude of 'new' particles as bound states of only three basic particles, the quarks. Regge trajectories have been alluded to above, and their relevance has dwindled to a point where, in my opinion, it is not necessary to discuss them. Consider them as an idea that at one time was appealing, but which did not work out.

Then the direction of thought changed radically. Instead of three quarks as building blocks, instead of thinking in terms of triangles one had to change to the family type structure described in Chapter 2. The drawing illustrates the point. What was a triangle became two straight lines, with the addition of a fourth quark.[g]

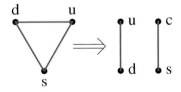

---

[g]The third line, with top and bottom quark, came later.

Now that kind of change of vision is hard to accept, in particular because the three-quark model worked so nicely. We now realize that this was because the quarks are all equivalent if one restricts oneself to gluon interactions. Thus apart from a relatively small difference in the masses of the quarks, resulting in small mass differences between the various bound states, there was not that much difference between those bound states. However, in a larger picture where also weak interactions play a role the view changes. The three quark picture became an accident, a part of a larger scheme, while before it was often viewed as a basic concept of Nature. The three-quark picture would have been a four-quark picture if the charmed quark had been much lighter, and it would have been a six-quark picture if also the top and bottom quark masses had been of the order of a few hundred MeV. One shudders to think what kind of particle zoo that would have given!

The change of view from triangle to two lines is historically not precisely what happened, the evolution was much more involved. It is impossible to say when this new vision took hold. But there were things of that nature, and this example is perhaps useful to illustrate the point.

It is interesting to note here another fact. Gell-Mann, when introducing the three quarks conforming to the triangle picture sketched above, made remarks that seemed at odds with this view. He mentioned lepton-quark symmetry, and as the leptons appeared in doublets (electron plus electron-neutrino and muon plus muon-neutrino) while the quarks seemed to form a triplet it was not clear what he meant. The Japanese physicist Hara, working at Caltech near Gell-Mann, introduced a fourth quark, and to a large extent produced the two quark doublet picture just discussed. Up to a point he produced the new picture. Despite his fabulous memory Gell-Mann does not really remember Hara, and I do not think that they had much interaction. Nonetheless, surely Hara found his inspiration in Gell-Mann's quark paper. Glashow noted Hara's work, and the fourth quark, named charmed quark by him and Bjorken, became part of Glashow's later work on the Standard Model (with Iliopoulos and Maiani). Then the picture became clear.

9

Particle Theory

9.1 Introduction

In 1948 quantum mechanics entered a new phase. Increasingly precise experimental results required new calculation methods, as the existing methods were hopelessly inadequate to deal with the complications of the theory. Richard Feynman came up with a new method that led to enormous simplifications. The method relied heavily on little drawings, now called Feynman diagrams. For a given situation one would draw a few of these diagrams, and then there were simple rules that provided the calculational answers in connection with them. As these diagrams are moreover very appealing intuitively they have become the universal tools of particle physics.

Historically, the work of Feynman was a tremendous step forward. In itself it did not really add to the theory, but it made working with it practical. It became simple to do calculations. The first domain conquered was the theory of photons and electrons, quantum electrodynamics (QED). That theory had its difficulties, but these difficulties could be overcome using a procedure called renormalization. That is discussed in this Chapter. Using Feynman's techniques this procedure becomes transparent.

Besides the interactions of photons and electrons there are other interactions, notably weak interactions. It took many decades to understand these forces. The renormalization procedure was not sufficient to eliminate all troubles. Progress came with the idea that new forces, new particles, with suitable interactions,

Richard Feynman (1918–1988). The most important contribution of Feynman, in my view, is his introduction of the diagram method named after him, and the theoretical tool, path integrals, that he developed. Truly wonderful work.

Part of the formal theory associated with those diagrams was published before, in French, by Stückelberg in the somewhat inaccessible journal, *Helvetica Physica Acta*. This including the idea that a positron may be viewed as an electron going backwards in time (this is basically the idea of crossing). It is unlikely that Feynman knew of that work, yet when he learned of it he dutifully acknowledged that in his papers. There are some anecdotes associated with that, not necessarily true.

On the evening of the day (in 1965) that Feynman celebrated his Nobel prize he received a telegram during the party: "Send back my notes, please", signed Stückelberg. According to my source (unpublished biography of Stückelberg by Ruth Wenger) the originator of the joke was Gell-Mann. I asked Gell-Mann if he had sent this telegram, but he denied that, adding that it was a nice idea.

When Feynman, after receiving his prize in Stockholm, gave a lecture at CERN, Geneva, he was afterwards introduced to Stückelberg. He asked Stückelberg: "Why did you not draw diagrams?" To which Stückelberg answered: "I had no draughtsman". Stückelberg, always the perfect gentleman and very conscious of his standing as a baron, apparently felt it below his dignity to draw those simple figures himself.

Feynman was a very charming person to talk to, and he was a gifted teacher. Well-known are his textbooks on physics, and he came very much in the public eye in connection with his part in the understanding of the Challenger disaster.

could be introduced such that the theory became manageable, i.e. renormalizable. The famous Higgs particle is one of those particles. From a mathematical playground this became reality when these hypothetical particles (except the Higgs particle) were actually discovered, and moreover were demonstrated to have the requisite properties. It is this work that was honored with the 1979 (Glashow, Salam and Weinberg) and 1999 ('t Hooft and Veltman) Nobel prizes. The work of Glashow, Salam and Weinberg concerned the construction of the actual model, while 't Hooft and Veltman elucidated the mathematical structure, showing that this model was renormalizable. By model we mean here a precise list of particles and their interactions. Without the simplifications due to Feynman's methods that progress would have been unthinkable. Not only in experimental physics but in theoretical physics as well the advance in techniques leads to new developments and insights.

9.2 Feynman Rules

Feynman rules are the main tools of the contemporary particle theorist. These rules incorporate the basic concepts of quantum mechanics; most importantly they can be represented in terms of drawings, diagrams, that have a strong intuitive appeal. A few basic concepts must be understood first to appreciate these drawings.

In Feynman diagrams particles are represented by lines, and interactions between particles by points where these lines join. Such an interaction point is called a vertex. The most obvious example is the interaction of electrons with photons. It is an interaction that we see literally almost permanently: the emission of light by electrons. In Feynman diagram language this interaction is represented in a very simple manner, see figure below. The electron, represented by a line with an arrow, shakes off a photon and moves on. The arrow is not there to indicate the direction of movement, but rather that of the flow of (negative) electric charge. Later on the meaning of the arrow will be changed slightly, but for now this will do.

The interaction of electrons with light has been well understood for a long time, and we have a precise quantitative understanding of the physics corresponding to this diagram. However, it must be understood that this simple diagram applies to many situations; the difference is in the possible initial and final configurations. That is typical for quantum mechanics: specify the initial and final configurations and then the theory provides the calculation of the probability for the process to happen.

Thus in practice each diagram must be supplemented with a precise specification of the initial and final state. Very often these initial states are particles coming from accelerators and the final states are the outgoing particles observed in detectors; in other words freely moving particles that collide or emerge from a collision. However, there are other situations. Concerning the above diagram the electron may initially be in a higher orbit in an atom,[a] and fall to a state of lower energy, a lower orbit, thereby emitting a photon. Another example is the emission of radio waves by an emitter. Electrons, moving back and forth in the antenna, shake off photons. In both cases the emerging photons are freely moving particles, but not the electrons, they are tied down in some way.

An important lesson that we can draw from this diagram is that particles can be created in an interaction. First the photon was not there, and some time later it came into existence. The opposite happens when a photon hits the eye: the photon is absorbed by an electron which then somehow leads to excitation of a nerve. The diagram corresponding to this process is shown in the figure below.

---

[a] Electrons can go into a higher orbit due to collisions between atoms or electrons or by absorbing light.

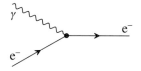

The difference is that here the photon is incoming, not outgoing. We "crossed" the photon line, i.e. we moved the line from outgoing to incoming. "Crossing" is an important property of Feynman diagrams: when moving a line from in to out or vice versa a new diagram results which corresponds to another possible process. This opens up interesting possibilities, especially if we apply crossing to other than photon lines, for example to electron lines. Let us consider the last diagram and apply crossing to the incoming electron line.

The result of this electron line crossing is another figure: a photon changes into an electron pair. According to the arrow one of its members has the charge moving in the opposite way, and we observe it as positive charge going out. So this is the rule: when a particle is outgoing, and the arrow points inwards we interpret that as the opposite charge. Thus this particle is now like an electron, except its charge is positive. It is called the positron, the antiparticle of the electron. So our crossing rule gets refined: crossing a line changes a particle into an antiparticle (and vice versa).

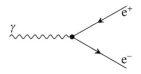

Positrons were experimentally discovered in the 1930s, and today antiparticles are an almost automatically accepted part of particle physics. Some particles are identical with their antiparticles: the photon is an example. Lines corresponding to such a particle carry no arrow and the particle has no charge.

Meanwhile there is another important element to understand. The interactions obey strictly the laws of energy and momentum conservation. An electron in an atom can emit a photon while dropping to a lower energy state. But a free electron cannot emit a photon. Consider an electron at rest, i.e. with zero momentum; it is then in its lowest energy state. If it were to emit a photon of finite energy then an electron with even less energy would be left behind, which is not possible. The same then holds also for a freely moving electron, which one could imagine to be an electron at rest as seen by a moving observer. Since the way processes go should not depend on the frame from which they are observed, especially not whether the observer is moving or not, it follows that if some process is not possible for one observer it should also not occur for any other.

Likewise a photon cannot change in mid-flight into an electron-positron pair, even if it is a high-energy photon. This can be understood by realizing that this high-energy photon appears as a photon of lower energy to another observer moving in the same direction as that photon. A photon always moves with the speed of light, and one can never catch up with it like in the case of a particle with mass; instead, when an observer races in front of the photon he will still see it coming with the speed of light, but it appears red-shifted, i.e. it is perceived as a photon of lower energy. If the observer moves fast enough, the photon energy can for this observer become less than needed to create an electron pair (whose energy at rest is twice the rest mass energy of one electron).

In other circumstances, where another object absorbs or adds some momentum or energy, photon conversion to an electron-positron pair can happen. In collisions with nuclei a high energy photon will in fact readily convert into an electron-positron pair. An observer moving in the same direction as the photon would see a photon of lower energy, but it would then from his point of view collide with a moving nucleus, and there is still enough energy for pair creation. An electron or positron moving through matter may likewise emit a photon, commonly called bremsstrahlung (literally brake-radiation). In Chapter 7 there is a bubble chamber picture of a neutrino event. In that picture one can see several electron-positron pairs, appearing like sea-gulls. These pairs are due to photons coming from bremsstrahlung by an earlier electron or positron. All of these processes involve an extra photon carrying momentum and energy from electron or positron to or from a nucleus; an example is shown in the figure. The fat line represents a nucleus.

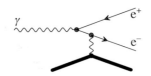

The next point is one of quantum mechanics. Particles can exist with 'inadmissible' energies, provided that this occurs only for a short time. The more inadmissible the energy, the shorter the duration. What we mean here by inadmissible is an energy different from the value that one must normally assign to a particle with a given momentum. For example, an electron at rest has an energy corresponding to its rest mass multiplied by the speed of light squared ($E = mc^2$). An electron with zero energy is hence not possible. Yet quantum mechanics allows the existence of zero energy electrons and even negative energy electrons, or of electrons with inadmissibly large energies (for example a very high energy electron at rest), provided this takes place only for short times. In particular, referring to the discussion above, a photon in flight can momentarily become an electron-positron pair, but very quickly the pair must recombine again into a photon. This possibility is shown in the figure below.

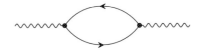

A particle in an inadmissible state of energy and/or momentum is called a **virtual** particle. Because the relation between energy and momentum is not that of a free particle ($E = mc^2$ for a particle of zero momentum) such a particle is said to be "off mass-shell". At this point we may recall Chapter 4 where the concept of mass shell was extensively discussed. Particles off mass-shell, virtual particles, are parts of diagrams, and we may even have some intuitive feeling about them, but we should never make the mistake of treating them as real particles. They occur as intermediate objects in a calculation, in a diagram, but they cannot be observed directly. They are like the photons in the two slit experiment. They move from light source to screen, and one may ask through which slit they pass. That, however, is a senseless question that can never be answered. We are not even sure if those photons actually go through any of the slits. That is the philosophy of quantum mechanics, and you better get used to it. Here we have diagrams and we can make calculations; it is like using wave theory to compute the interference pattern on the screen. But be careful not to think too much of the virtual particles as real objects. Still, within limits, it is helpful to think of them as a variant of the particles that they represent. Consider a virtual particle as a sort of calculational help. It makes you understand processes in a more intuitive way, and that is the path that we shall take.

Keeping the above remarks in mind, and using the language of virtual particles, it follows that a photon, for a small fraction of time, can become a virtual electron-positron pair. This is actually of some consequence if we let another photon cross the path of the first one. On the level of diagrams new possibilities arise. If the second photon catches the first one in a dissociated state, it could be absorbed by one of the two virtual particles, to be emitted again by the other. The figure below shows the diagram.

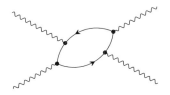

As we know precisely the quantitative value of the photon-electron coupling (the previous diagrams), and also know the quantum-mechanical behaviour of these particles, the probability of the process can be calculated. It would be observed experimentally as the scattering of light by light. You might say that it is still possible to have interactions with virtual particles. The effect is quite small, so you cannot observe it by crossing the beams of two flashlights. Nonetheless, it has been observed. The classical (Maxwell) theory of radiation does not allow such a process. It is a purely quantum-mechanical effect.

The effect just described is somewhat similar to the so-called tunneling effect, well known to students of quantum mechanics. A particle may cross an energy barrier even if it has not enough energy to go over the top. An electron could cross a mountain even if it had not enough energy to get to the top. It may "tunnel" through. The tunnel, however, should not be too long; the probability for this to happen goes down very quickly as a larger distance must be covered. Here again there is the question. If the electron is initially on one side of the tunnel and finally at the other side, it seems only natural to say that "it has passed through the mountain". Such a statement is however beyond the limits of quantum mechanics. There is no way to establish if the electron actually ever was halfway in the mountain. The electron is there a virtual electron. The moment that you try to locate it (analogous to establishing through which slit a photon passes) the effect disappears. In an intuitive sense the electron passed through the mountain, and you may use that picture to devise experiments, such as sending other electrons from other directions, having them influence one another inside the mountain. In other words, in some sense interactions between virtual particles are quite possible. What you observe, however, are the initial and final configurations, never the intermediate virtual particles.

One more effect must be discussed, namely interference. Light interferes with itself, and this is a property of particles as well. The way this works is that for a given situation there may be

Willis Lamb (1913). In a splendid series of experiments Lamb found a discrepancy in the spectrum of hydrogen, as compared to the theoretical predictions of quantum theory. The discrepancy, now called the Lamb shift, was reported at a conference on Shelter Island (near Long Island) in June 1947. Kramers, successor to Lorentz in Leiden, the Netherlands, had been aware of the possibility of such effects, and lectured about his insights. The participants then recognized that the effect was due to higher order effects of quantum field theory, not taken into account up to then because people did not know how to handle the infinities of that theory. In his lecture Kramers also came up with the idea of renormalizability. The classic example is that of a small ball moving through water; a thin layer of water attaches to that ball and moves with it, thereby effectively increasing its mass. Likewise, the electron mass supposedly derives partly from the energy contained in the electric field around that electron. Kramers suggested that one must clearly separate the "bare" mass (the mass of the electron not including the contribution of the field) and the physical mass, that what you see experimentally. The bare mass is in fact not observable, and one simply chooses it in such a way that after addition of the field energy the observed mass value comes out. That the calculation of the field energy produces infinity is regrettable, but by choosing a bare mass of minus that same infinity (plus something extra) the correct experimental result can be reproduced. That is the kind of thing theorists do: sweeping infinities under the rug, smuggling them away. Ugly as it is, the theorists present at Shelter Island (including Feynman) followed Kramers' suggestion, and produced a calculation of the Lamb shift that agreed with the experimental results of Lamb.

more than one way to go from a given initial state to a given final state. These different possibilities may interfere, either constructively or destructively. That is certainly something in which elementary particles seem to differ from billiard balls or cannon balls. In actual fact, the laws of quantum mechanics apply equally well to macroscopic objects; the point is that for the latter the effects become too small to be observable. Imagine a machine gun firing point-like bullets at two slits; the interference pattern on the screen would be incredibly small, the distance between the top in the middle and the adjacent peaks would be something like 10^{-37} m. That is much, much smaller than the size of a nucleus!

In calculations with particles the theorist draws as many diagrams as applicable (i.e. diagrams with the same initial and final configuration), writes down the corresponding mathematical expressions and sums them up. The different possibilities may add up or subtract, i.e. interfere. Only from this sum total can the probability of the process happening be calculated (effectively by squaring it). For example, to compute light-by-light scattering one must consider six diagrams, and combine their contributions. The figure below shows the possibilities.

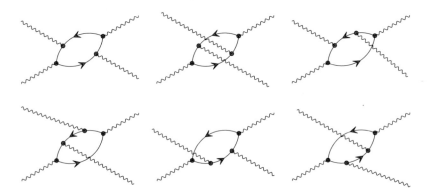

All of these diagrams correspond to contributions of possibly different sign, and these contributions interfere. After taking all these contributions together the result must be squared and that is then a real probability. Never, but never could one say that one or the other possibility as represented by the different diagrams was what actually happened. Again, that would be like asking through which slit the photon passed.

9.3 Infinities

Where life becomes difficult is implicit in these diagrams. Not only must they be summed over all the different configurations, but over all the different energy-momentum values of the virtual particles as well. Consider for example again the temporary transformation of a photon into an electron-positron pair. The virtual electron or positron of this pair can have an infinite range of possible energies, including also negative energies. For example, the electron may be very energetic, while the positron would have very negative energy. The particles are then very far off mass-shell. The total energy must of course be equal to the energy of the photon, energy conservation being strictly enforced by Nature.

In calculating a process one must sum over all the possibilities. One must take all virtual configurations into account, no matter how much off mass-shell. That leads often to a hard calculation. Moreover, sometimes the summation gives an infinite result. It is a question of magnitude of contributions. If the configurations with an electron-positron pair of very high energy (very high negative energy for one of them and very high positive energy for the other) keep on contributing as much as configurations with low energy electron-positron pairs, then there is simply no end to the summation. The central question then is to what extent configurations containing virtual particles very far off mass-shell keep on contributing in a sizable manner as compared to contributions of configurations very nearly on mass-shell. One may put it in the following way. Normally contributions are smaller, damped, as the particles are more off mass-shell. The crucial thing is the amount of damping. If there is no or too little damping one is in trouble. **This, in a nutshell, is the problem of infinities in quantum field theory**.

Let us try to give an example for the case of sound. Imagine yourself standing in a crowd, hearing the conversations around you. Normally conversations held by people at some distance will not bother you very much because the volume of sound decreases with the distance. Thus there is a damping factor associated with sound generated at some distance. Imagine now that there would be no such damping factor, that a conversation far away would be heard by you as strongly as a conversation nearby. That would be horrible. You would hear a conversation between two Chinese in Beijing and two Russians in Moscow as strongly as a nearby discussion. Clearly, you would go mad. The noise would be unbearable.

Actually, in a large room filled with people, the sound volume would go down with distance, but there is the opposite effect of there being more people in a larger circle around you. That would more or less overcome the distance damping effect. In a large room full with people the total amount of noise may be very large indeed, and you may have to shout to carry on a conversation yourself! And if the room were infinitely large you might go deaf.

Several factors affect the occurrence of the infinities mentioned. To begin with, the more a particle is away from its mass shell the shorter is the time it is allowed to exist in that state. Consequently there is normally a damping factor associated with the occurrence of any virtual particle. This damping is stronger as the particle is more virtual. Furthermore, the damping is also a function of the intrinsic properties of the particle (more about that below). Another factor is the behaviour of the vertices, i.e. of the coupling, as a function of the energies of the particles involved. By and large these couplings have no strong energy dependence, although there are exceptions.

A difficult point is the behaviour of virtual particles as function of their intrinsic properties. The main property in this respect is the "spin" of the particle. One of the very surprising discoveries in the domain of quantum physics was the discovery that particles have an intrinsic angular momentum, as if they were spinning around an axis. For a body of some size, like a billiard ball, that is easy to imagine. But for a particle that for all we know has no size that is very hard to imagine. Yet it is there, and every elementary particle has a very definite spin as this intrinsic angular momentum is called (it may be zero). When particles are created or absorbed the interaction is always such that angular momentum is conserved. If a spinning particle enters an interaction then the angular momentum is preserved throughout, and it appears in the final state either in the form of other spinning particles, or else through non-spinning particles that revolve around each other, or both. All this is quite complicated, but fortunately we need only a few facts related to this. Spin is measured in terms of a specific basic unit, and spin is always a multiple of $\frac{1}{2}$ in terms of that unit.

As it happens, no elementary particle observed to date is of the spin zero variety. The so far hypothetical Higgs particle has spin zero. Most particles have spin $\frac{1}{2}$, and the remainder have spin 1, except for the graviton (the particle responsible for gravitational interactions similarly to the photon in electromagnetism) that has spin 2. Here now is the important property relevant to our discussion about virtual particles: as their spin becomes higher, virtual particles are less damped at higher energy. Particles of spin 1 are barely damped at high energy in their contributions to a virtual process. Particles of spin 2 are even worse: the quantum theory of gravitation is in a very poor shape. Quantum field theory for particles of spin 1 (with the exception of the photon) was not part of our understanding of Nature up to 1971. No one knew how to handle the virtual contributions. They invariably led to infinite sums.

Even if the photon has spin 1, and thus has not much of a damping factor associated with it, there is still effective damping due to the way that different diagrams tend to compensate each other. As a consequence the theory of electrons interacting with photons became manageable, using the renormalization technique discussed below. In 1948 Feynman, Tomonaga and Schwinger worked out the theory for which they received the Nobel prize in 1965. However, the weak interactions, involving other spin 1 particles, remained intractable.

What changed the situation for weak interactions was the discovery that the worst effects in individual diagrams can, theoretically, be cured by introducing new interactions and particles (and hence new diagrams) in such a way that in the sum total the bad parts cancel out. Thus new particles were introduced into the theory. That will be discussed at length later on. For the photon a similar mechanism, involving cancellations between different diagrams but without the introduction of new particles, was, as mentioned above, partly understood since 1948. Using the method of renormalization quantum electrodynamics produced finite numerical results that could be compared with experiment. One of the most important results is the magnitude of the magnetic moment of the electron. Any charged particle with spin usually has a magnetic moment, which one might consider as a consequence of the spinning charge. The predictions of quantum electrodynamics for the magnitude of this magnetic moment have been verified to a truly fantastic degree of accuracy. But before discussing this we must first fill in some gaps, and explain about perturbation theory.

9.4 Perturbation Theory

As we have pointed out, one must sum up all possibilities when considering any process. That includes summing over all energy/momentum distributions of the virtual particles. However, also additional emissions/absorptions of virtual particles must be taken into account. The figure shows an example: the virtual electron emits a photon which is absorbed by the virtual positron (or the other way around).

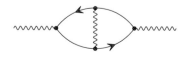

Horace R. (Dick) Crane (1907). Crane, a professor at the University of Michigan, discovered a very sensitive method to measure the magnetic moment of the electron. In certain units, the value of this magnetic moment (usually called g) is 2 if no radiative corrections are included. When an electron moves through a magnetic field its magnetic moment (pointing in the same direction as its spin) will not change relative to the direction of motion along the trajectory if indeed this g is exactly 2. However, any deviation of the magnetic moment from that value will cause the electron spin to rotate. By measuring the amount of rotation an accurate measurement of the difference between the actual value and the value 2 can be measured. In this way one measures pure and simple the quantum corrections that cause the magnetic moment to have a value different from 2. This method is called the $g - 2$ method. The theoretical calculations and the various experiments on this magnetic moment have occupied many physicists in the course of time.

Crane is a modest man who is not given to advocate his own achievements. Things being what they are in this business he did not always get the proper credit for his wonderful idea. He did get the US presidential medal, a very high US distinction.

On retirement he decided to help the Ann Arbor Hands-on museum (an initiative of Cynthia Yao, wife of a theorist at the University of Michigan). He managed to keep most items exhibited in working order, which is a must for survival of any museum of this type. How Crane did it is a mystery to me, because visitors, often including whole bus loads from schools, are not particularly careful.

In his spare time Crane grows orchids.

Richard Garwin (1928, left) and **Valentine Telegdi** (1922). These two physicists did pioneering experiments (each a different one, with other collaborators) on parity violation after the theoretical analysis of Lee and Yang. They showed that the muons in the decay of the pion ($\pi \rightarrow \mu +$ neutrino) were polarized. This implies parity violation in pion decay.

Concerning $g - 2$, after the original proposal of Crane, physicists were quick to realize that the method could be very suitable to measure the anomalous magnetic moment of the muon. The muon decays in a way that depends on its spin orientation (also this indicates parity violation), and by recording the decay products the direction of the spin (and thus the magnetic moment) could be measured relatively simply. At high energy the muon lives long enough to make it traverse a substantial distance.

On the theoretical side the necessary theory (not the quantum corrections) was first formulated by Ken Case, a theorist of the University of Michigan. A subsequent classic paper by Bargmann, Michel and Telegdi became the standard concerning the treatment of spin.

At CERN a group of physicists mounted the first muon $g - 2$ experiment, in 1959. In 1961 a result was published (Charpak, Farley, Garwin, Muller, Sens, Telegdi, Zichichi) with an accuracy of 1.9%, agreeing perfectly with theory. Since then, using muon storage rings at CERN and Brookhaven, this accuracy has improved greatly.

Telegdi is a long-time friend of mine who helped me enormously with this book. Of Hungarian origin, he became a stateless person at the end of World War II and traveled around with a self-made passport.

Here we have another complication of quantum theory: there is no end to this chain. One can exchange one photon, two photons, whatever number, and to boot any of these photons can momentarily become an electron-positron pair, etc. It looks hopeless. How can one calculate all these diagrams?

As luck has it, there is in many cases no need to consider all of these possibilities. The reason is that there is a factor associated with any vertex and that factor, at least for quantum electrodynamics, is quite small: the electric charge. The emission or absorption of a photon by an electron (or positron) is proportional to the electric charge of the electron. Indeed, if the electron had no charge it would not interact with the electromagnetic field. For this reason, a diagram as shown above with an additional photon exchanged between electron and positron gives a contribution that is down by a factor e^2, where $-e$ is the electric charge of the electron. In practice there are some additional factors, and the relevant dimensionless quantity is what physicists call the fine-structure constant, $\alpha = e^2/4\pi\hbar c$. Numerically $\alpha \approx \frac{1}{137}$, so that a diagram with an extra photon exchange indicates a contribution of the order of 1% as compared to that of the diagram without that photon exchange. So if we restrict ourselves for a given process to diagrams with the least number of vertices we may expect an answer that is accurate to 1%. And if that were not enough we can include diagrams with two more vertices and get an accuracy of 0.01% (i.e., 1 part in 10^4).

Here we see a fact of field theory: it is **perturbation theory**. Rarely, in fact never, can we compute things exactly but we can approximate them to any desired precision. That is of course true assuming we can find our way though the maze of summations (over energy/momentum distributions) that arises when considering a diagram with many virtual particles. The calculation of the magnetic moment of the electron is in practice perhaps the most advanced example. The electron, possessing spin, has like any rotating charged object a magnetic moment. In other words, the electron has not only a charge, but it also acts as a tiny magnet as well. That part of the interaction of an electron with the electromagnetic field is also subject to quantum corrections, and the figure below shows the lowest order diagram and a next order (in α) diagram.

In suitable units the magnetic moment of the electron, disregarding quantum corrections, is 1 (in another context units are often chosen such that it is 2). The second and higher order contributions alter that magnetic moment by a tiny amount; to give an idea about the accuracy achieved we quote here the theoretical result for this anomalous magnetic moment (including fourth and sixth order contributions as well):

$$a_e = 0.5 \left(\frac{\alpha}{\pi}\right) - 0.328478965 \left(\frac{\alpha}{\pi}\right)^2 + 1.181241456 \left(\frac{\alpha}{\pi}\right)^3 - 1.4 \left(\frac{\alpha}{\pi}\right)^4$$

$$= 0.001159652201$$

as compared to the experimental value 0.001159652188. Note that $\alpha/\pi \approx 0.00232$. The error margins in both the theoretical and experimental values are of the order of the difference between the values quoted here. In other words, the agreement is excellent. Also the anomalous magnetic moment of the muon has been computed with great precision: $a_\mu = 0.00116591849$ (there is an uncertainty of 70 in the last four decimals), to be compared with the most recent experimental value $a_\mu = 0.00116592030$ (uncertainty 80). The sophistication involved in both theory and experiment is mind boggling. The calculation of the coefficient of α^3 has taken some 20 years, involving some 72 diagrams, while the calculation of the α^4 term (891 diagrams) has been done mainly by numerical approximation methods, using up years of super-computer time. Any experiment achieving an accuracy of one part in a thousand is already difficult, let alone the experiment relevant here, having an accuracy of order of one part in 10^6. The most spectacular experiment for the electron is based on measurements performed on a single electron, caught in an electromagnetic trap (Dehmelt, Nobel prize 1989). For the muon things are slightly different, because it happens that with this accuracy, for the muon, diagrams involving a W or a Z^0 or even quarks must be included. That makes it even more interesting. At the time of this writing the latest measurement of the magnetic moment of the muon, quoted above, was done at Brookhaven using a muon storage ring.

Here a remark concerning the way theorists talk about these things. They usually classify the subsequent orders of perturbation theory by means of loops. The lowest order diagram is called a tree diagram (no loop), the next order diagrams have one-loop, the next order two loops etc. The figures below show examples of a two-loop and a three-loop diagram.

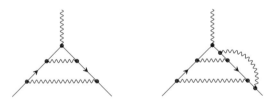

The indeed amazing agreement between theory and experiment, involving these very complicated quantum effects, must be seen as a strong support for the theoretical insights as well as for the validity of perturbation theory. Many theorists would have liked a formulation of the theory not involving approximations, but so far perturbation theory is all we have. In certain instances one has been able to sum up the contributions of some classes of diagrams to all orders, but we do not have any general non-perturbative version of the theory. This is a fact of life. Let us be happy with the notion that we know at least the basic ingredients that form the basis of the quantum mechanical calculations.

In my own scientific life this wonderful agreement between theory and experiment has played an important role. It made me deeply conscious of the fact that diagrammatic methods and perturbation theory worked very well, and this stimulated me to continue using these techniques even in the dark times in the middle sixties when false gods were dominating particle theory. Now, of course, with the Standard Model we can apply these methods all over the place.

9.5 Renormalizability

As indicated earlier, there is a problem related to the non-convergence of the summations over all possible distributions of energy/momentum of the virtual particles. In certain cases these sums do not converge, i.e. the result is infinite. That stopped progress for quite some time, until, in about 1948, the idea of **renormalization**, due to Kramers, solved the problem at least on a practical level. The subject theory was quantum electrodynamics, and it was noted that the infinities occurred only in some well defined instances. For example, in the calculation of the magnetic moment of the electron discussed above they did not occur. But the reader will realize that these very same diagrams, which alter the magnetic properties of the electron, will equally well alter the electric charge of the electron, as they simply affect the way a photon interacts with the electron. That change of the electric charge turns out to be infinite. Here then is the big idea: the electric charge as actually observed is the sum total of the basic electric charge (as occurring in the tree diagram) plus the contributions of all higher order diagrams. But we have no idea how large that charge (the basic charge) is without the corrections due to these higher order diagrams.[b] So, let us choose the value of the basic charge such that the total charge comes out equal to the experimentally observed value. In other words, we give the basic

---

[b]Such corrections are called radiative corrections.

charge a value that has an infinite part as well, but opposite to that part of the higher order corrections, making the sum come out equal to the observed value! Speaking of dirty tricks!

This trick, hard to swallow at first, works very well indeed. The basic fact is that the infinities occur only in conjunction with the free parameters of the theory. A free parameter is a parameter for which there is no theoretical prediction. Any theory has some of these parameters. For example, Newton's theory of gravitation has the gravitational constant. It fixes the strength of the gravitational force. You can determine it by working out one case, for example the orbit of the earth. Basically, the constant is determined from experiment. There is no theory that says how big Newton's gravitational constant should be. It is a free parameter, to be determined from experiment.

The electric charge of an electron is also a free parameter. It is an input to the theory, not something that we can compute. Another such parameter is the mass of the electron. It is not known from any basic principle, and its value must be obtained by measurement. That gives us an opportunity to hide an infinity. Experimentally one observes the basic value of the parameter, but theoretically, what is observed is some input value plus the contribution of many diagrams. The important thing here is that what we observe experimentally includes contributions of all kinds of diagrams. At that point one says: whatever the contribution of diagrams, it goes together with the basic value and the only thing we know is the combination, observed experimentally. Thus if the diagrams give an infinite contribution let us make the basic value also infinite, but with the opposite sign so that the combination comes out to the experimentally observed result. Nonsense minus nonsense gives something ok.

This scheme for getting rid of infinities is called renormalization. It is by itself far from satisfactory. No one thinks that the basic quantities are actually infinite. Rather we believe that the theory is imperfect, but that this imperfection can be isolated and at least for the moment swept under the rug. The miracle is that for quantum electrodynamics all infinities can be absorbed into the available free parameters. So, apart from these infinite corrections to free parameters, everything else (such as the magnetic moment of the electron quoted above) is finite, and insofar as checked, agrees well with experiment.

So here we are. We have a theory imperfect on several counts. First, the theory is only perturbative. Second, infinities occur, even if they can be isolated and hidden. In spite of these imperfections, all this leaves us with a scheme that makes accurate predictions that can be compared with experimental results.

There are also theories such that infinities occur not only in conjunction with free parameters. Such theories cannot make solid predictions. They are called **non-renormalizable**. For a long time theories involving vector particles (spin 1) other than the photon were thought to be of that type, and as such useless. This picture has changed, and we now also have renormalizable theories involving spin 1 particles. These theories are called **gauge theories**.[c] Strictly speaking the name gauge theory refers not to renormalizability but to some mathematical property, which in the case of spin 1 particles leads to a renormalizable theory. There are gauge theories that are not renormalizable, gravitation (involving a spin 2 particle) being one of them. The name **Yang-Mills** theories (named after the inventors, C. N. Yang and R. Mills) refers more narrowly to gauge theories with particles of spin 1.

As a matter of fact, almost all interactions seen in experiments are of the renormalizable type, gravitation being the exception. Quantum effects in gravitation have so far not been understood. That casts a shadow on that theory and its consequences, like for example black holes.

---

[c]Historically the name gauge was introduced in another context, namely gravitation. It referred at the time to a transformation, a gauge transformation, that changed the length scale. There are similar transformations in the theories that we discuss, but they do not scale length. Nonetheless, the name has stuck.

9.6 Weak Interactions

Weak interactions constitute a different type of interactions, one that does not produce long-range forces such as those in electro-dynamics. A photon has zero mass, and can hence have arbitrarily low energy. For this reason a virtual photon of zero energy and small momentum is only a tiny bit "off mass-shell", and little damping is associated with the exchange of such a virtual photon. It is this type of zero energy, low momentum photons that are responsible for long-range electromagnetic interactions. For the same reason the graviton will also give rise to a long-range force. These, however, exhaust the list of long-range forces that we experience in daily life. The weak interactions have a very short range; let us discuss them in some detail.

Weak interactions made their entry into physics through the discovery, by Becquerel in 1896, of β-radioactivity. Experimentally and theoretically it took a really very long time before these interactions were understood, even on a purely phenomenological level. Since it is not our purpose to present here the history of the subject, we shall straightaway describe things as they are under-stood today.

Consider the most fundamental nuclear β-decay, that of a neutron into a proton and an electron (plus an antineutrino). The neutron contains two down quarks and one up quark (denoted by d and u respectively), the proton one d and two u quarks. As a first step, one of the d quarks in the neutron decays into a u quark and a negatively charged vector boson, denoted by W^-. The figure below shows the diagram representing this decay.

It contains one of the basic vertices of weak interactions. The associated coupling constant ("weak charge"), usually denoted by

g, is somewhat larger than the corresponding one for electromagnetism. Experiment shows that $\alpha_w = g^2/4\pi\hbar c = 1/32$. At this point we have a notational problem, because all particles in this reaction are charged (in terms of a unit such that the charge of the electron is -1, the charges are $-\frac{1}{3}$, $+\frac{2}{3}$ and -1 for d, u and W^-, respectively), and the arrow no longer represents the flow of negative electric charge. Instead it will be used to distinguish particles and antiparticles, where an arrow pointing opposite to the flow of energy indicates an antiparticle. Here there is a choice: is the W^- a particle or an antiparticle? Historically, there was the regrettable mistake of having defined the charge of the electron as negative. We shall not do that here for the W, and define the W^- to be the antiparticle. That is why the arrow in the W-line points inwards.

The W is very massive (80.3 GeV, as compared to the proton mass, 0.938 GeV), and given the low mass of the d quark (about 10 MeV = 0.010 GeV) it must be very virtual (way off mass-shell) in the actual process of d decay. In a second step it must quickly transform into an electron and an antineutrino, an interaction which is another basic vertex of the theory. The figure shows the complete diagram for d decay. Let us note here that an antineutrino is not the same as a neutrino, despite the fact that the neutrino has zero electrical charge.

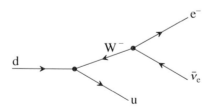

As noted before, the existence of a negatively charged W^- (which we will take to be an antiparticle) implies the existence of a particle with opposite charge, the W^+.

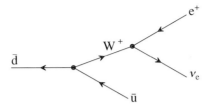

This W^+ would, for example, be involved in the decay of an antineutron into an antiproton, a positron (anti-electron) and a neutrino; that reaction is simply the same reaction as the one discussed above, with all particles replaced by their antiparticles (reversal of all the arrows).

The W has spin 1 and its interactions generally lead, due to the absence of damping at high energies, to a non-renormalizable theory as we discussed before. We shall discuss this point in a systematic way, showing how the situation can be salvaged, and how new particles and interactions must be introduced in order to achieve a tolerable high-energy behaviour.

Let us point out that we have quietly introduced another property of Feynman diagrams. First there was the crossing property discussed before. Lines may be moved from in to out or vice versa, and that gives new possible processes. The point introduced above is this: if in a given diagram all arrows are reversed then another process results, and this new process can also occur in nature. For example, the very first diagram in this Chapter, an electron emitting a photon, when treated this way becomes a positron emitting a photon, see figure below. Once more: the arrow has nothing to do with the movement of the particle; the fact that the arrow points opposite to the movement of the particle (the flow of energy) means that we are dealing with an antiparticle.

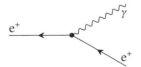

9.7 Compton Scattering

Exploring high-energy behaviour can conveniently be done by considering certain simple processes. One must estimate the energy dependence of a process where particles of very high energy are scattered. The simplest and most important example is Compton scattering, that is the scattering of a photon incident on an electron. In lowest order only electromagnetic interactions are of relevance here, and the figure shows the two diagrams that contribute at the lowest level.

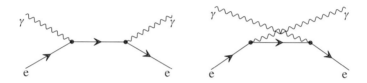

Strictly speaking much of what follows below is true only if photons have non-zero mass, but we shall ignore this subtle point for the sake of simplicity.

The central issue is the behaviour of the theory as dependent on the energy of the particles. It is a necessary property that the theoretical behaviour must not be such that the probability of the process to occur increases if the energy of the incoming particle (in this case the photon) increases. Then that probability would for sufficiently high energy become larger than one, which is nonsense and certainly not acceptable for a physical theory.

We now state a general property of the diagrams shown. Each of these diagrams alone would produce a bad theory. Taking into account only one of the two diagrams, the probability for the process to happen will increase indefinitely if the energy of the photon increases. And that is unacceptable. However, in this case the two diagrams combined produce an acceptable result, the probability is constant if the energy of the photon goes up. The two diagrams compensate each other. This is the wonderful thing: diagrams may compensate each other's bad behaviour. And indeed, that is what does the trick for quantum electrodynamics.

The high-energy behaviour of diagrams as shown above and similar diagrams with different particles can be guessed as follows. We are not explaining anything here, just stating the rules. An incoming or outgoing photon (vector particle, spin 1) contributes a factor proportional to the energy E of that photon. A virtual photon contributes no energy dependence, i.e. it must be counted as a constant. A virtual electron, or generally a virtual spin $\frac{1}{2}$ particle behaves as $1/E$. An incoming or outgoing electron (spin $\frac{1}{2}$ particle) must be counted as $\sqrt{E}$. A virtual scalar particle (spin 0) must be counted as $1/E^2$, an incoming or outgoing spin 0 particle contributes a constant. It must be noted that in special cases the energy dependence might be different from the dependence that one would deduce by counting with these rules; whether or not it does depends on details of the actual couplings which sometimes compensate the aforementioned energy dependence related to the magnitude of the spin. A case in point, one that we shall meet in the following, is the coupling of a vector boson to two real (i.e. not virtual) particles. In the case of gauge theories the vertices are always such that spin effects are neutralized in that instance, i.e. for a (possibly virtual) vector boson coupling to two real particles one can ignore the spin effects for this vector boson. An example is the decay of the down quark, corresponding to a diagram that we will show again.

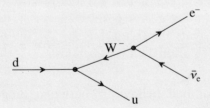

Here the virtual W^- is coupled to real particles on both ends. In that case the energy dependence relating to that virtual W is as that for a scalar particle, i.e. as $1/E^2$. For the readers' convenience, we summarize the various factors given above in the following table.

spin	in/out	virtual	ends[*]
0	1	$1/E^2$	$1/E^2$
$\frac{1}{2}$	$\sqrt{E}$	$1/E$	$1/E$
1	E	1	$1/E^2$

[*]Behaviour of virtual particles with real particles attached at both ends and coupled according to a gauge theory

In a renormalizable theory the probability of a process to occur must, as a function of energy, either decrease or at worst tend to a constant value. Even on a purely intuitive level a probability increasing indefinitely as the energy of the incident particle increases is hard to accept. Note that the probabilities (cross sections) are obtained, as mentioned earlier, by squaring the contributions of the diagrams. Counting the expected behaviour for the diagrams shown above for Compton scattering we arrive at a result increasing with energy, in fact as E^2 (so the cross section would go as E^4). This then is unacceptable. However, the second diagram shows also a leading dependence proportional to E^2 but it turns out that it has the opposite sign, and the sum of the two actually behaves as a constant. A somewhat simplistic explanation for this difference in sign is that the intermediate (virtual) electron in the first diagram has a large positive energy, in the second a large negative energy. Thus the factor $1/E$ for the intermediate electron has the opposite sign for the two diagrams.

Cancellations of bad behaviour between diagrams is the idea behind gauge theories, of which quantum electrodynamics is the simplest example. Individual diagrams give unacceptable energy behaviour, but everything is arranged in such a way that in the end the bad behaviour cancels. It is a complicated game of cancellations, requiring from time to time the introduction of new particles. Experiment must then verify the existence of those new hypothetical particles having the desired properties. Let us see if we can make this work for weak interactions that involve charged vector bosons.

9.8 Neutral Vector Bosons

In the following we shall ignore electromagnetic interactions, giving rise to diagrams containing a virtual photon in some of the situations discussed below. They are not essential to the reasoning.

Let us first examine W^+ scattering off an electron. The corresponding lowest order diagram is the first one in the figure below. There is a virtual neutrino mediating this process. The behaviour at high energy is bad.

> The high energy behaviour guessed by power counting as specified above is bad, namely as E^2 (E for each of the W's, $\sqrt{E}$ for each of the electrons and $1/E$ for the intermediate neutrino). As the W's are connected to vertices of which one of the particles (the neutrino) is virtual there are no special compensating effects.

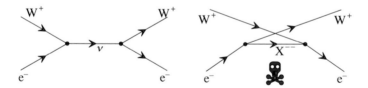

Recalling the case of Compton scattering we might think that the situation can be cured by another process, the one shown in the second diagram in the figure above. Since now the incoming electron emits a positively charged W, the intermediate particle (named X^{--} in the diagram) cannot be a neutrino because charge conservation (a law that holds rigorously) requires the intermediate particle to have charge -2. But no such particle is known. The diagram does not exist. That is why we put a skull and bones below it. What now?

274

Steven Weinberg (1933) and **Martinus Veltman** (1931). Weinberg and I do not see eye to eye on certain issues. The picture above, taken at the occasion of the 1999 Nobel week, shows Weinberg offering me some explanations that I found difficult to swallow. In 1967 he wrote his most famous paper (for which he was awarded the 1979 physics Nobel prize, together with Glashow and Salam), but up to 1971, at which time the mathematical consistency of his model became clear (after the Amsterdam conference) it was largely ignored. It did not help that in this paper only one experimental consequence was mentioned, followed by the remark that it should not be taken very seriously. Indeed, nobody did. Weinberg refers to the period 1967–1971 as the period that his paper lay dormant. It is now, I believe, the most cited paper in particle physics, followed by the paper of Kobayashi and Maskawa (see Chapter 3).

In 1972 at a conference at Fermilab, Ben Lee, reporting on theory in a session entitled "Perspectives on theory of weak interactions", pulled Weinberg's paper out of obscurity. Ben Lee's talk was very important, as it explained many of the facets of gauge theories to a large audience; not long after that neutral currents (a consequence of the existence of the Z^0) were established by a neutrino experiment at CERN using the gigantic French bubble chamber Gargamelle.

Weinberg's paper contains one of the ingredients that made the Standard Model what it is today. He invoked Higgs forces that we now know to be necessary for mathematical consistency, adding them to the model proposed earlier by Glashow.

Sheldon Glashow (1932) and **Gerardus 't Hooft** (1946). Glashow is certainly one of the main contributors to the Standard Model. His 1963 paper specified the interactions between leptons and vector bosons (W and Z^0), thereby introducing the Z^0 and moreover the mixing of the photon with the Z^0. That paper was Weinberg's starting point. The idea of lepton-hadron symmetry of Gell-Mann (see the end of Chapter 8) was implemented by the Japanese physicist Hara, who introduced a fourth quark next to the up, down and strange quarks proposed by Gell-Mann. That new quark is now called the charmed quark. Glashow, together with Iliopoulos and Maiani, spelled out the interaction of these four quarks with the vector bosons (without a Higgs though). Among others that work explained the hitherto mysterious absence of certain decays of the K-mesons. One speaks of the GIM mechanism.

Also reactions of neutrinos without production of muons or electrons (neutral currents) were discussed. In 1973 neutral current type events were seen in the French bubble chamber Gargamelle, convincing many physicists of the correctness of that part of the Standard Model.

In 1968 I convinced myself of the importance of gauge theories, and made substantial inroads in this complicated mathematical subject. At some point 't Hooft became my PhD student and he then did his work that completed the mathematical understanding of those theories. He delivered a splendid piece of work, and at the time I was very happy with that and proudly introduced him to the physics community at the 1971 Amsterdam conference. Being one of the organizers I was left at liberty to arrange a session of that conference.

The solution is to introduce another vector boson, this time one without electric charge. We assume that it couples to the charged vector bosons and the electrons and a new diagram of the third type is then possible, see below. The coupling of this new neutral particle to the charged vector bosons and the electrons is taken such that the high energy behaviour of the new diagram cancels the bad behaviour of the first diagram above.

The vertex must behave like E, and given that the intermediate vector boson is coupled on both ends to real particles we have indeed the required behaviour, E^2 (E for each of the charged W's, $\sqrt{E}$ for each of the electrons, E for the three vector boson vertex and $1/E^2$ for the intermediate vector boson). Choosing the right sign and magnitude for the coupling constants in the various vertices one may achieve cancellation, and the sum of the two diagrams behaves as a constant at large energies.

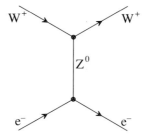

The price to pay is the totally ad hoc introduction of a new particle, a neutral vector boson. But here starts the triumph of gauge theories: a neutral vector boson with the required couplings has indeed been observed. It is commonly called the Z^0. Its mass, 91.187 GeV, is slightly higher than that of the charged W's.

9.9 Charmed Quarks

Another interaction observed is the decay of the Λ. This neutral particle is very much like the neutron, except that it is heavier

(1116 MeV versus 940 MeV for the neutron). The Λ particle has one d, one u and one s (strange) quark, from which you may guess that the s quark is about 200 MeV heavier than the d quark. The Λ decays in various ways, there being more energy available, but one of its decay modes is quite analogous to neutron decay, namely decay into a proton, electron and antineutrino. That decay can then be understood as a decay of the s quark into an u quark and a W^-, with, as in the case of neutron decay (or rather d quark decay) a subsequent rapid decay of this W^- into an electron and an antineutrino. See figure below.

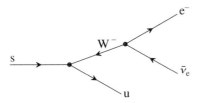

It is found that the coupling constant associated with the $s-u-W$ vertex is smaller by a factor of about $\frac{1}{4}$ as compared to the coupling constant of the $d-u-W$ vertex. This factor has been interpreted as the tangent of some angle, now called the Cabibbo angle θ_c; in that view, now generally accepted, the d quark decay has a factor $\cos\theta_c$ in addition to the coupling constant g, while s decay has a corresponding factor $\sin\theta_c$. It is interesting to see what must be done in order to achieve the proper high-energy behaviour for the scattering of a W from an s quark through the process as shown in the first diagram in the figure below. The

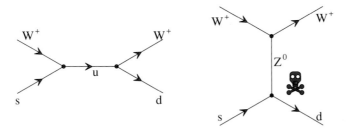

quark-W vertices of both Λ and neutron decay are involved here. Given that the quarks are spin $\frac{1}{2}$ particles, just like the electron, we guess just as before a bad high-energy behaviour for this process.

Again a compensating diagram must be found, and naturally the first thing that comes to mind is to try the same trick as found earlier, namely to introduce a diagram involving a Z^0, as shown in the figure above. This however fails. The reason is that there occurs here a new vertex, the $s-Z^0-d$ vertex, which, as far as experiment is concerned, does not exist. In the language of particle theorists this was called "the absence of strangeness changing neutral currents" (the s quark has strangeness, the d has not). How to repair this?

Well, the solution was to postulate yet another particle, one with properties close to that of the u quark, but with suitably chosen couplings to the W's. The figure shows the construction; it is completely analogous to the diagram with the intermediate u quark given above.

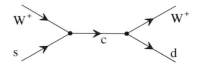

The new quark is called a "charmed" quark.

The coupling constant of the $s-c-W$ vertex is assumed to be like the one in neutron decay, i.e. with a factor $\cos\theta_c$, while for the $c-d-W$ vertex it is taken to be $-\sin\theta_c$. Due to this extra minus sign the diagrams almost cancel at high energy, and their sum behaves neatly like a constant.

As if asked for, experimenters discovered this c quark, or rather, since quarks are never seen singly, discovered particles that contained these c quarks. The mass of the c quark is of the order of 1500 MeV, much heavier than the masses of the d quark

(7.5 MeV), the u quark (5 MeV) or the s quark (200 MeV). Even so the c quark is much lighter than the b (bottom, 5000 MeV) and t (top, 175,000 MeV) quarks found since then. Incidentally, these masses are not well established, in particular not the lighter ones, because quarks never appear singly, and there are substantial energies related to the binding mechanism peculiar to quark interactions. By necessity the quark masses must be derived from the experimentally accessible particles, quark composites, and that always requires elaborate arguments.

The discovery of the c quark was the second major victory for the gauge idea. Its couplings were found to be precisely as given above, involving the appropriate factors $\cos \theta_c$ and $\sin \theta_c$. But the story does not end here.

9.10 The Higgs Particle

We now turn to processes involving vector bosons only. Of course, our considerations are here of a purely hypothetical nature, since one cannot observe in the laboratory any of the scattering processes discussed here. Vector bosons have very short lifetimes and hence cannot in practice be used as projectiles or targets. Anyway, we shall now consider vector bosons scattering from each other, in particular $W^+ - W^+$ scattering. There are two diagrams contributing to this process (see figure below), and the behaviour at high energy is really bad. That is because compared to previous cases there are now only spin 1 particles (as compared to the occurrence of spin $\frac{1}{2}$ particles in the previous diagrams), and higher spin makes for worse behaviour. Drastic remedies are needed.

Recall that the $W - W - Z^0$ vertex has an energy dependence (a factor E). The powercounting gives a behaviour as E^4: a factor E for any of the four external W's, a factor of $1/E^2$ for the intermediate Z^0 (it is coupled to vertices without other virtual particles), and a factor E^2 coming from the two vertices.

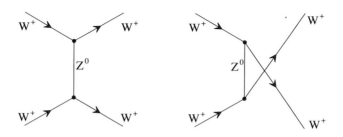

The worst part can be cured by introducing a totally new type of diagram, namely a direct four-W interaction as shown in the figure.

By carefully adjusting the coupling constant associated with this new vertex an almost complete cancellation can be achieved. However, there is still trouble remaining, although much reduced.

The vertex itself is not energy dependent, and the four W's give a factor E^4, so it has at least the required behaviour. The E^4 part can be cancelled. This one new diagram is not enough. A part behaving as E^2 remains. Rather, since its dimensions have to be the same as those of the E^4 part, it is a behaviour of the form M^2E^2, or $M_0^2E^2$ where M and M_0 are the masses of the W and Z^0 bosons, respectively. These masses are the only available parameters with the dimension of an energy. How to compensate the remaining E^2 part?

The solution is to postulate yet another particle, called the Higgs boson H. It is to be a spinless neutral particle, coupling to the W's with a strength so chosen as to produce the required compensation. The figure shows the two possible diagrams.

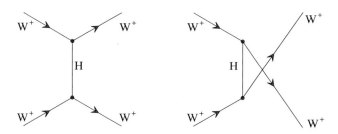

It turns out that the Higgs particle must be coupled to the vector bosons with a strength proportional to the mass of the particle that it couples to. This peculiar feature, typical for all the couplings of the Higgs particle, raises many interesting questions.

9.11 General Higgs Couplings

Is this the end of the story? Not quite. There remain many little problems of the nature sketched before, but it would carry us too far to enter into a detailed discussion here. Suffice it to say that the Higgs particle must also be coupled to the neutral vector boson (the Z^0) and to the quarks etc. as well. In short, it must be coupled to **any** particle having a mass. Moreover, the coupling must always be proportional to the mass of the particle to which it is coupled.

To date the Higgs particle has not been observed experimentally. Unfortunately the theory has nothing to say about its mass, except that it should not be too high (less than, say, 1000 GeV), or else its compensating actions set in too late. The present experimental lower limit for the Higgs mass is roughly 100 GeV. The new collider being built at CERN (the LHC, colliding protons each with an energy of 7000 GeV) might give us information on this Higgs particle. It will not be easy to produce Higgs particles, because the proton contains only u and d quarks, and these, because of their low masses, couple only weakly to this Higgs particle. Higher order processes, involving virtual (heavy) W's and Z^0's are needed to produce this particle.

The demonstration that all bad energy behaviour can be made to vanish with only those particles discussed above, and no others is usually referred to as the proof that this theory is renormalizable. On reading the previous discussion one may easily have the impression that there is no end to new hypothetical particles that must be introduced. But no, this is it! The Higgs particle is the last one needed.

It is perhaps necessary to state explicitly to what extent the discussion above reflects the historical development. We have sketched a theory involving many particles, with their interactions so orchestrated and tuned as to have a renormalizable theory. The result is a theory possessing a high degree of symmetry. The historical development was quite the opposite of that suggested by our treatment. The symmetry was discovered and investigated some 20 years before its consequence, a renormalizable theory, was finally understood.

9.12 Speculations

Because this Higgs particle seems so intimately connected to the masses of all elementary particles, it is tempting to think that somehow the Higgs particle is responsible for these masses. Up to now we have no clue as to where masses come from: they are just free parameters fixed by experiment. It requires no great imagination to suppose that the Higgs particle might have something to do with gravitation, and indeed, theoretical models suggest a strong involvement of the Higgs particle in the structure of the Universe, otherwise thought to be shaped by gravitation. Some theorists believe that the Higgs particle does not really exist, but that it somehow mimics a much more complicated reality, involving gravitation in a fundamental way.

These are very exciting and interesting questions and speculations. We are looking forward to LHC experiments, noting that so far theorists have not been able to come up with any credible theory that answers all or some of these questions, including questions concerning the magnitude of the masses, the Cabibbo angle, the existence of all these quarks, the grouping of these particles into families, and so on. There is clearly so much that we do not know! Even so, we have certainly made enormous advances in understanding the structure of the interactions between the elementary particles.

9.13 ρ-Parameter

In addition to W^+–W^+ scattering other processes may be considered, such as for example W^-–Z^0 scattering, or Z^0–Z^0 scattering. All these processes can be made to have decent high energy behaviour, but to cure all of them using only one Higgs particle requires a relation between the charged and neutral vector boson masses, usually expressed in the form $\rho = 1$, with $\rho = M^2/(M_0^2 \cos^2 \theta_w)$. Higher order quantum corrections slightly modify this relation. In this equation M and M_0 are the masses of the W (80.3 GeV) and the Z^0 (91.2 GeV) respectively. To explain the angle θ_w appearing here would require a detailed discussion about the interplay of weak and electromagnetic interactions, due to the fact that wherever the Z^0 couples to charged particles on both ends, the photon can take its role. Experimentally one finds $\sin^2 \theta \approx 0.2315$, and we conclude this discussion with the observation that ρ comes out to the predicted value so that there is no need to have more than one Higgs particle.

It is interesting to note that the higher order corrections to the equation $\rho = 1$ involve among others, the mass of the top quark in a most peculiar way: the correction becomes bigger as the top quark mass is heavier. Here we have a quantum effect that increases if the intermediate state is energywise further away! Many years before the top quark was actually observed the measured magnitude of the quantum corrections was used to predict the top quark mass. That prediction agrees quite well with the experimental value.

The reason that the radiative correction grows with the top mass is a very typical consequence of a gauge theory structure. The top quark has a function in the scheme, for if it is not there certain diagrams grow in an intolerable way. So, if you try to eliminate the top quark from the theory (by making it very heavy) you are left with an infinity. The figure shows the relevant diagrams, which concern momentary dissociation of the W^+ and Z^0 into a quark-antiquark pair. Such diagrams are called self-energy diagrams. The effect we discuss involves the first diagram minus $\cos^2 \theta_w$ times the sum of the second and third diagram (since the first diagram gives a correction to M^2, the other two to M_0^2). The top quark is now essential; without the top quark only the second diagram would be there, and this diagram all by itself gives an infinity.

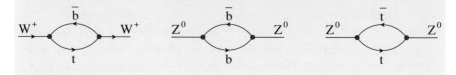

10

Finding the Higgs

The Standard Model has been very successful. All the detailed couplings and particles needed for the cancellations between the diagrams have been found, and the more complicated quantum mechanical corrections that can be calculated theoretically agree with the observed data. The most spectacular of these is the quantum correction to the ρ parameter and the mass of the top quark; it might be helpful to remind the reader in simple terms of this most remarkable incident. We will be repeating some of the arguments of Chapter 9, but it is from a slightly different point of view, and also such that it is not necessary to understand all of the sometimes difficult arguments of that Chapter.

First of all there is the issue of the number of Higgs particles. While the theoretical difficulties that the Higgs particle(s) must cure are well defined, it is quite possible to cure those problems using one, two, or in fact any number of Higgs particles. So here is the first question: how many Higgs particles are there? Here we have some idea about the answer.

The theory by itself has nothing to say about the values of the masses of the vector bosons. They must be established by measuring them. However, it so happens that if all theoretical problems mentioned in the previous Chapter are to be solved using one and only one Higgs particle then the ratio of the mass of the charged vector bosons (W^+ or W^-) to the mass of the neutral one (Z^0) must have a very specific value. Thus by measuring the masses of the vector bosons we have an indication of the possible number of Higgs particles. Here experiment tells us

Peter Higgs (1929). This is the man whose name is associated with the big mystery of the Standard Model: the Higgs particle. He developed his work after an important piece of work by Anderson, who investigated the penetration of electromagnetic fields in a superconductor. He found that they penetrated only over a small distance, and he produced a theoretical understanding. The same mechanism was then used by Higgs (and Brout and Englert) to make a theory of photons with mass, thus with a limited range for the associated force. That became an ingredient of the Standard Model, not for the photons (that have zero mass), but for the 'homologues', the vector bosons W and Z. That turned out to be just what was needed to make the model mathematically viable (renormalizable).

I met Higgs for the first time at a summer school in Edinburgh in 1959, where he was part of the organization, in particular he had the key to the wine cellar. Cabibbo and I, among others, profited greatly from his gracious understanding of our needs.

Polkinghorne, a professor at that same summer school, writes in a book (*Rochester Roundabout*): "Higgs was a competent theorist, but of no great distinction," while in that same book he writes about X (admittedly a very good physicist): "X is a very deep thinker with a marked reluctance to publish his ideas," and of the same X with respect to Higgs's idea: "Perhaps one of the most surprising aspects of the story is that this idea did not occur to the acute and fertile mind of X himself." In 1979 Polkinghorne became an Anglican priest, instantly becoming the best physicist among Anglican priests. Recently he received the enormous Templeton prize. I think it was for something indeed not that easy: bridging the gap between sense and nonsense.

Robert Brout (1928, left) and **François Englert** (1932). These two must be credited, together with Higgs, for introducing the Higgs system. They are not well known to the general public, the name of Higgs alone has stuck in association with the subject.

Brout and Englert were perhaps the first to suspect that their work would make vector boson theories renormalizable. Englert said so in a discussion remark at a talk by Weinberg at the 1967 Solvay conference. Weinberg did have a handwritten version of his 1967 paper with him, in which the same is said. However, I do not think that any of them had any inkling about the complexities of the theory.

I feel a bit guilty with respect to Brout and Englert, because in 1971 I heard about Higgs's work, but only later of their work, and I thus did not cite them in the beginning. That was one of the reasons that they did not become as well-known as Higgs. They got full recognition in 1997, when they were awarded (together with Higgs) the High-Energy and Particle Physics prize of the European Physical Society.

I met Brout in Utrecht around 1958. He impressed us all, but I did not meet him for a long time after that because he worked mainly in another field of physics, more related to the domain in which Anderson was active.

At a dinner where Englert was also present I proposed a conjecture based on the statistics of one person (myself), namely that being born in the summer, preferably June, is the best with respect to intelligence. Englert, born in November, replied by saying that he was a Jew, and did not need this. Then he laughed so hard that I started to be worried for his life. For your information: my conjecture holds on the average for Nobel prize winners; Einstein, however, was born in March.

that the answer is that the values of the masses are indeed precisely such that one Higgs particle is enough to do the job.

However, there are subtleties here. The mass of a vector boson such as the Z^0 is affected by quantum corrections (often called radiative corrections). The fact that a vector boson such as the Z^0 may, for a short time, split into a pair of particles of different mass, changes slightly the value measured for the Z^0 mass. The particles that may intervene here are any of those that the Z^0 is coupled to, and that includes the top and bottom quark. The figure below shows the two possibilities.

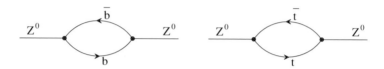

The same story holds for the charged vector bosons. The measured value of the mass of the W^+ is also influenced by the occurrence of virtual pairs, but they occur differently as compared to the Z^0 case. In fact, there is only one possible diagram. The reader may observe that this is so because of conservation of charge. See the next figure, and remember that the antibottom quark has a charge of $+\frac{1}{3}$ while the top quark has a charge of $+\frac{2}{3}$.

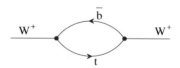

The masses of the W^+ and W^- are the same, and so are the corrections due to quantum effects. However, the corrections to the Z^0 mass are not equal to those of the W^+ mass, and the ratio of charged and neutral vector boson masses changes slightly. This rather small effect has been evaluated theoretically and measured experimentally, and as its value depends on the masses of top and bottom quark, the measurement can be used to determine the top mass (the mass of the bottom quark is of course known since its discovery in the seventies). This then led to a prediction for the top mass, and indeed the top was found with precisely that value for its mass. It in fact helped the experimenters to find the top in 1995, since they had at least some idea about its mass before they started looking for it. At that moment the Nobel committee started worrying about what became the 1999 Nobel prize. Predicting and experimentally confirming the mass of an elementary particle is the sort of thing they look for.

It is here that we resume our discussion of the Higgs particle. It too may influence the vector boson mass measurements, and possibly also the ratio of the masses. See the figure below. There is a strange new type of diagram that ought to tickle the imagination of the reader.

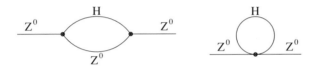

As the Higgs couples with different strength to the Z^0 and the W^+ the mass ratio is indeed affected, although very much less so than through the corrections due to bottom and top quark. So, the prediction for the top quark mass is unsure since we do not know the mass of the Higgs particle and thereby the magnitude of the correction (except that its magnitude is quite small). In practice this led to an uncertainty of about 5% in the prediction of the top quark mass. Once however the top quark was discovered and its mass measured, the mass of the Higgs could be estimated from this very small effect. The prediction for the Higgs mass from this is not very precise, and today stands at somewhere above 110 GeV with a very large error margin.

So, what is the situation? Most likely there is only one Higgs (if any!) and there is a vague prediction for its mass. However, there is trouble brewing and it is not at all sure that the Higgs actually exists. Here are the complicating issues.

Introducing the Higgs with certain couplings to the known particles (all of which have to be verified when the Higgs is actually found) leads also to the necessity of coupling the Higgs to itself. So two Higgs bosons may attract each other, etc. This turns out to have some really surprising consequences. There may be bound states of Higgs particles, depending on the strength of the Higgs self-coupling. The reader may recall the important feature of a bound state, namely that it is a state of negative energy as compared to the non-bound state. This is obvious if one realizes that it costs energy to tear the bound state apart. For example, in a hydrogen atom the electron is bound to the proton with an energy of 13.6 eV. You need 13.6 eV of energy to pull the electron from a hydrogen atom. Thus in a bound state there is some amount of negative energy, binding energy, in addition to the usual mass-energy. The total energy of one hydrogen atom is equal to the sum of electron and proton mass minus the binding energy.

Thus a bound state of Higgs particles involves negative energy. It now happens that it is possible to have bound states of two, three, etc., Higgs particles, and there is a bound state of an infinite number of Higgs particles whose binding energy is actually larger than the sum of all the Higgs masses! Thus the total energy of that state is negative, and if you start with nothing then you can create energy by making such a bound state. This is a most curious and disturbing fact, because it is obvious that such a state (it actually has an infinite spatial extension) would be created immediately in the beginning of our universe. But such a bound state cannot go undetected. Having a system of Higgs particles all over the universe is something that would be sensed by gravitation, and calculation reveals that such a system would lead to a curved universe with the size of a football. Theoretically it must be cured in a most horrid way: one assumes that the universe was initially curved in a negative sense and in precisely the same amount before this Higgs bound state came along. The result then would be a flat universe. Quite unbelievable, unless there is a principle that forces these two a priori unrelated curvatures to be the same. Recent observations by astronomers have shown that the universe is really very flat and even the expected curvature due to the masses of galaxies etc. has not been seen. There is a big puzzle here. Evidently there is some relation between the Higgs system and gravitation. How strange.

All this indicates that there is more to the Higgs than meets the eye, and one may well expect to see something quite different from the simple picture of a particle of some 150 GeV with certain interactions with the known particles and itself. Even so the Higgs particle has a task to fulfill: it must be such that it cancels out certain unwanted effects in scattering processes. There are certain things that must be there with certainty, and therefore the hunt for the Higgs is not open-ended.

There is an important theoretical fact: the strength of the self coupling of the Higgs is related to the mass of the Higgs. So, a heavy Higgs couples stronger to itself than a light Higgs. If the Higgs is sufficiently heavy these self-couplings become so large that perturbation theory breaks down. Diagrams with an ever increasing number of self-couplings are not smaller than diagrams with no or only a few such couplings. Under those circumstances the theorists cannot make precise quantitative predictions.

Here is then the state of affairs. Higgs or Higgs related effects become large and visible (but actually unpredictable in precise magnitude) if the Higgs itself has a mass exceeding 500 GeV. The ominous fact that already now the Higgs, from an experimental point of view, seems heavier than any other particle we know seems to point in this direction. Of course, if the Higgs is that heavy the prediction from the vector boson mass ratio becomes worthless also, and cannot be used to predict the Higgs mass. So the Higgs mass prediction from present data may be a joke. We must go hunt for the thing itself, or inspect closely those situations where it has a job to do, cancelling undesirable behaviour for example in vector boson scattering. The theory ends here. We need help. Experiments must clear up this mess.

There are a few light points here. It is virtually sure that the Large Hadron Collider at CERN will come into operation in the first decade of this century. Very likely that will establish at least some as yet unanswered questions. The Higgs itself may actually be discovered if its mass is not too high (somewhere below 400 GeV). This machine does not cover, however, the complete spectrum of manifestations of the Higgs particle. Much more in that sense can be expected from a high-energy electron-positron collider, of which there are some on the drawing board. They may actually measure the Higgs self-interactions. But there is no question about it: we may well run into something totally unexpected!

Quantum Chromodynamics

11.1 Introduction

The theory of interactions of photons with charged particles is commonly called quantum electrodynamics, abbreviated as QED. The theory of interactions of gluons and quarks, considerably more complicated, is named quantum chromodynamics, or QCD. The quark-gluon interactions are responsible for the binding of the up and down quarks in proton and neutron, and generally for all those bound states discussed in Chapter 8 on the particle zoo.

The interaction of photons with electrons can be well described using perturbation theory. This means that one can restrict oneself to simple diagrams with only a few vertices. Every vertex implies a factor $\alpha \approx 1/137$, and diagrams with many vertices are numerically speaking very small. This is also true for the interaction of W's and Z^0 with quarks and leptons (and with each other) because the associated coupling constant $\alpha_w \approx 1/40$ is still quite small. However, the situation in QCD is very different: the coupling of the gluons to the quarks is large, of the order 1. In other words, diagrams with many vertices are just as important as diagrams with only a few vertices. Hence perturbation theory is no longer possible, and as a consequence the theorist is at a loss. Even so, much work has been done and much understanding has been gained. But no one has so far been able to calculate the mass of the proton or the pion even though we think that we can understand these objects as bound quark states.

Claude Bouchiat (1932, left), **John Iliopoulos** (1940, middle) and **Philippe Meyer** (1925). These three theorists played an important role in the development of the Standard Model. The part of the Standard Model containing only leptons is not really free of difficulties. There is a nasty, somewhat hidden infinity, and it is the virtue of these three physicists that they found the solution to this problem. They showed that the quark sector has the same sort of infinity, with the opposite sign, so that in the whole no infinity is left. This is only true if the quarks are threefold degenerate, i.e. have three colors, and their work was a very strong endorsement of the colored quark picture as shown in Chapter 2. It made us all believe in quantum chromodynamics, with its colors and gluons.

Bouchiat, as French as they come, was a student of Louis Michel. In the dark ages in France, when theory was stifled by de Broglie and his successors, Michel and Bouchiat were exceptions. Here an anecdote told to me by Michel.

Leprince-Ringuet, director of the Ecole Polytechnique, decided that his laboratory needed a theorist, largely for window dressing. He hired Michel, and at certain occasions, showing around visitors, he would open the door of Michel's office and announce: "Voila notre théoreticien" (Here our theorist). After that he would slam the door.

Iliopoulos was also involved in other parts of the Standard Model. He was one of the authors (with Glashow and Maiani) of the celebrated GIM paper that showed how to integrate the charm quark into that model.

Meyer was educated in the US. In that way he overcame the dark ages mentioned above. With Bouchiat he started a particle theory group in Orsay. The international summer institute organized by this group was of great importance, as this created the opportunity to interact with leading theorists. I attended these summer sessions for some 10 years, and I used them also to introduce my students in the community. At some time the group moved to the Ecole Normale, Paris, where they started a theory group. The hard part was to get space, which is at a premium in Paris.

There is another very important difference between the interactions of photons and those of gluons. There is no basic photon-photon interaction, the photon is neutral and does not couple directly to photons, because photons couple only to charged particles. With gluons the situation is different. There are eight different gluons, and they do couple to each other. To get an idea of what this means imagine that there would exist another version of the photon but with charge. Let us call that photon γ_c. The original, neutral photon (γ) would couple to this charged photon γ_c because the photon couples to any particle with charge. So imagine the situation around a nucleus: first there is the usual long range electric field, the Coulomb field, essentially made up from photons. Secondly there would be another long range field, like the Coulomb field, but charged and due to the charged photon. The trouble is now that these two fields interact, which produces a very complicated situation. That is the kind of situation that arises with gluons. Each gluon by itself would produce a Coulomb-like field around any quark. But these eight possible fields interact with each other, tying things up in a truly complicated manner.

Nonetheless, despite all this a good deal is known about the interactions of quarks and gluons, partly theoretically, and partly because of what one sees experimentally. Three important concepts, not unrelated, play an important role. The first is confinement, the second is asymptotic freedom and the third is scaling.

11.2 Confinement

The long-range interactions between gluons are theoretically unmanageable. It is not only a complicated mess, there are also new infinities popping up. It seems impossible to have long range gluon fields although the gluons, like the photon, are massless. An infinite amount of energy would be associated with these self-interacting long range fields. The solution is to assume that any object that can exist as a physical particle must be color neutral so that there are no long range gluon fields. That is like saying that, if there were similar problems for the photon, all particles must be neutral in order not to have a long range electric field. Thus all bound states of quarks must have a color combination such that they are essentially "white". For example, one may have a red quark bound to an anti(red quark). Or a certain combination of three quarks of different color, red, green and blue, which also produces no long range gluon fields. That, incidentally, is the beauty of the color code assignment: the property that white light can be obtained as a combination of red, green and blue light holds in this sense also for QCD. It is not immediately obvious how such a color combination has in the end no effect, but it would require equations to explain it. Protons and neutrons are supposedly bound states of three quarks of different color. The gluons emitted by this combination compensate each other when at a distance large compared to the distance between the quarks.

A consequence of this is that one cannot remove just one quark from a proton. The result would be a quark and a system of two quarks, neither of them color neutral, thus forbidden. It would need an infinite amount of energy to effect this separation. This then is confinement: the three quarks are confined to a small region near one another.

Confinement has been understood up to a point, but there exists no rigorous theoretical proof. Bound states of quarks have been discussed in Chapter 8, and we may recall the main point. The picture that particle physicists have is that one sees the gluons as some form of glue. As the quarks separate there remains a string of glue between the two, and moreover the quarks may rotate around one another. It should be added though that all this is somewhat wishful thinking, not solidly supported by the experimental facts. The image should not be taken too literally. The picture below shows a sketch of the situation of a bound state with some associated bound states of higher energy with additional rotation of the quarks around one another.

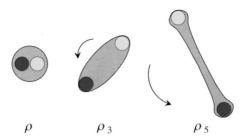

$$\rho \qquad \rho_3 \qquad \rho_5$$

Note that in this picture yellow represents the color antiblue, i.e. white without blue. Yellow contains equal amounts of red and green and no blue. White of course is equal amounts of red, green and blue.

Concerning the mass of the up/down quark bound states it is clear that the major amount of energy is in the gluon field, at least for the light quarks u, d and s.

11.3 Asymptotic Freedom

As we mentioned in the introduction perturbation theory cannot be used for the quark-gluon interactions because the coupling constant is too big. Here is the good news: the coupling constant is not a constant, but depends on the energy. And at high energy the coupling constant becomes small, smaller as the energy increases. This is called asymptotic freedom. The theoretical work by Hugh Politzer, David Gross and Frank Wilczek on this subject convinced the particle physics community of the validity of quantum chromodynamics as the theory of strong interactions.

How can we understand this? It is a quite complicated issue and at this point the reader might just accept the statement for what it is. But let us make an attempt to explain this, or rather tell where it comes from. It is a matter of radiative corrections.

Consider the coupling of an up quark to a gluon, see figure below.

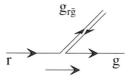

Here a red up quark emits a red–antigreen gluon and becomes a green up quark. There is a coupling constant associated with this, called α_{qcd}, and for low energies it is of the order 1.

In addition to this diagram there may be much more complicated diagrams in which the final products are the same (that is a red–antigreen gluon and a green quark). An example is shown in the figure below. Whenever an experimenter observes this transition he will have no idea whether he sees something corresponding to the first diagram or the last diagram. In fact, he sees the sum of the two. The last diagram has more vertices than the first, but because the associated coupling constant is nearly one that does not mean it is small. This is very different for the interaction of photons with electrons; a comparable diagram there would be smaller by at least a factor $\alpha^2 \approx (1/137)^2 \approx 1/18000$, thus well below the 0.01 percent level.

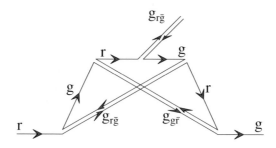

The above figure shows an example of a possible diagram. The double lines represent gluons, and the overall process is a red quark (on the left) becoming a green quark (right), emitting a red-antigreen gluon (top). We indicated colors by means of letters because actually drawing in colors makes it very messy. This

James Bjorken (1934). In 1967 he was a theorist working at SLAC and the experimenters confronted him with the rather amazing facts found in their electron-proton scattering experiment. While up to 1967 such experiments showed a decreasing scattering cross section with increasing energy of the electron (conform the idea that charge was homogenously smeared out within the proton), at the SLAC machine energy this was no more true. Somewhat earlier Bjorken had proposed scaling, a property that relates scattering processes at different incident electron energies. That indeed was verified to be correct. Then, in 1967, Bjorken suggested that scaling might be true if there would be elementary constituents inside the proton. An important breakthrough was due to Feynman who made Bjorken's somewhat highbrow arguments explicit by using an extremely simple picture: point particles inside the proton. Feynman called them partons, but today we call them quarks.

When I went to SLAC in 1963 Bjorken was working with Drell on a book that for quite some time was the standard textbook for elementary particle physics. At the time we were all housed in the workshop of SLAC, where they were constructing the machine, separated from all the activity by only a few low partitions. The noise was incredible. Bjorken and Drell were discussing their book across from my office, and I often wondered if there was real communication going on. I believe that the origin of some weird convention that they used for the metric (never mind what it is) is to be found there. I myself often escaped to the computer center, writing my program Schoonschip that could be used to do complicated symbolic manipulations. It helped me very much in evaluating Feynman diagrams.

diagram is only one of a large multitude of possible diagrams, and they must be included as well. It seems hopelessly difficult to compute anything here. But as it happens some very smart people have nonetheless been able to extract some facts here. It turns out that these extra diagrams do depend on the energy of the initial and final quarks and the gluon. Including all possible diagrams a very simple statement could be made: the energy dependence is such that increasing the energy has precisely the same effect as making the coupling constant smaller. From this evolves the following statement: the theory of quantum chromodynamics at energies of, say, 10 GeV is equivalent to the very same theory at 1 GeV but with a smaller coupling constant. In other words, going to sufficiently high energy the coupling constant of the equivalent theory at low energy may be sufficiently small to allow per-turbative calculations (amounting to considering only diagrams with a few vertices). And indeed, today quite successful pertur-bative calculations are done in the domain of very high energy quark-gluon interactions.

11.4 Scaling

The development sketched above started with the experimental observation of electron-proton scattering at SLAC, the Stanford Linear Accelerator Centre (1990 Nobel prize to the experimenters Friedman, Kendall and Taylor). There very high energy electrons were used, and the observed results showed very interesting regu-larities, called scaling by the experts. This scaling behaviour, sug-gested by Bjorken, a SLAC theorist, amounts to the fact that the results observed for some electron energy are quite simply related to those observed at a different energy. Then Bjorken suggested that scaling could result if there were elementary constituents inside the proton. He did all that using the rather formal math-ematical language of field theory. Feynman succeeded in interpret-ing these results in a lucid and amazingly simple manner, and we will try to outline the principal idea.

Consider an electron scattering off a proton. In the first instance this is ordinary Coulomb scattering, see figure.

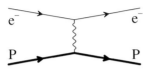 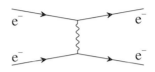

Indeed, at very low energies it behaves precisely like that, i.e. just like electron-electron scattering. This latter process, electron-electron scattering, is well described by this single diagram up to very high energies (the different sign of the electron charge is of no consequence in this process, only the magnitude counts). More complicated diagrams need not to be considered, as they are quite small, due to the smallness of the e.m. coupling constant $\alpha \approx 1/137$. But a proton, as we know now, is a very complicated system, consisting of three quarks embedded in a sea of gluons. If the photon exchanged is of low energy it sees the proton as a whole, but if the energy increases the photon penetrates into the proton and scatters off the individual quarks. That is where it becomes complicated. The gluon mass surrounding and interacting with the quarks becomes part of the process.

It is here that asymptotic freedom comes to the rescue. If the energy is sufficiently large the coupling constant of the quark with the gluons becomes small, and the quark starts behaving as a free quark. At that point the process becomes precisely like the electron-electron scattering process (except that the charge of the quark is different, namely $+\frac{2}{3}$ for an up-quark or $-\frac{1}{3}$ for a down quark, which can easily be taken into account). So, if the photon gives a sufficiently large kick to the quark it will at least for some small distance behave as a free quark.

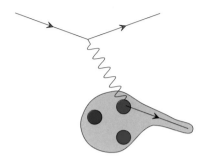

Later on the recoiling quark will be caught up as it must remain confined, but by that time the scattering process is over. The proton will be ripped apart by the recoiling quark in a complicated manner, of which the details are way too involved to understand quantitatively. For example, a gluon may momentarily become a quark–antiquark pair of the appropriate color, of which one may team up with the recoiling quark to make a colorless object such as a π^+ that can escape confinement.

The figure below shows a proton breaking up into a neutron and a π^+. The recoiling up-quark is for example a blue up-quark, while a $d\bar{d}$ pair (due to the dissociation of a gluon) could have the colors blue and antiblue (= yellow). The antiblue $\bar{d}$ combines with the recoiling blue up quark to make a colorless π^+.

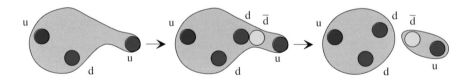

So this is the idea: at high energies quarks behave like free particles. For this reason the SLAC experiments as interpreted by Bjorken and Feynman helped tremendously to establish the reality of the quark picture. Before that quarks were rather abstract things, mathematical entities, explaining the symmetry observed in the particle zoo. The SLAC experiments, properly interpreted, made them into real things. Subsequently, the discovery of asymptotic freedom for quantum chromodynamics made QCD the appropriate theory describing the interactions of quarks and gluons. Of all theories that was the only one giving rise to asymptotic freedom. At the same time that theory, with its behaviour at low energies (where the interaction becomes really strong) provided at least a qualitative understanding of confinement.

$$\widehat{12}$$

Epilogue

Chapter 9, about particle theory, is perhaps the most important Chapter in this book, but perhaps for many also the most difficult. In that Chapter the problem of the forces between the particles was tackled. The delicate balancing mechanism of the Standard Model has been discussed: starting with some process the existence of new particles, needed to create new possibilities, was explained. All this amounts to a balancing of forces in such a way that no process shows undesirable features. Studying processes involving the W led to the introduction of the Z^0; at a later stage, in a similar manner the Higgs particle was introduced to balance out a remaining bad piece in $W-W$ scattering. That "bad piece" is actually open to experiment: in certain processes involving the production of W pairs the behaviour of the process depending on the energy may be studied experimentally. In this way one will be able to establish experimentally the need for the Higgs, and if that particle is not seen, show how Nature is going to solve that problem. That is the exciting part of things to come: perhaps Nature has another way of doing things, and perhaps that other way gives us some deeper insight in the many other problems that we have.

At this time the Standard Model is well confirmed experimentally. At the big electron-positron collider LEP at CERN, Geneva many precision measurements on many reactions have been made, all agreeing with the theory to within experimental errors. Fermilab at Chicago has found the top quark with a mass precisely as calculated using these precision measurements as an input.

Aerial picture of CERN, Geneva. In the foreground the runway of the airport of Geneva. The dashed line indicates the frontier between Switzerland and France. The circles indicate where underground the tunnels housing the accelerators are located. The large circle, with a diameter of 8.5 km did contain LEP, the Large Electron-Positron collider; it has been removed and now the LHC (Large Hadron Collider) is being constructed there. That is where our hopes for this decade are.

CERN is located at the point where the small circle touches the large one. In the tunnel corresponding to the small circle (diameter 2 km) is the SPS (Super Proton Synchrotron). At CERN there are yet smaller machines, the PS (Proton Synchrotron), diameter 200 m and the ISR (Intersecting Storage Rings for 30 GeV protons), equally large. The latter two are not indicated.

CERN is managed by a directorate headed by the Director General. When I came to CERN in 1961 that was Victor Weisskopf (1908–2002, shown above). He was in my view the best we had. He initiated the ISR and the SPS. He was a member of the Pontifical Academy; when his statements on arms control were used by the Pope he noted dryly that it was remarkable that the Pope quoted a Viennese Jew.

A council made up from representatives of all the participating (European) countries meets several times during a year and discusses policy, finances, etc. An advising body to the council is the SPC (Scientific Policy Committee) whose members are elected on the basis of their physics merits. In the beginning Heisenberg was a member of this committee. I was member in the second part of the seventies, pushing for LEP, helped among others by Cabibbo, Telegdi and Dalitz.

However, we should also be well aware of the limitations of our understanding.

Perhaps the greatest mystery of them all is the remarkable three-family structure of quarks and leptons. No one has found any explanation for this structure. We are reasonably sure that there are no more than three families. The clearest evidence for this comes from the observation of the decay of the Z^0. It can decay only into particles together of mass less than the Z^0 mass, but if there is a fourth family with another massless or near massless neutrino then the Z^0 could decay into a neutrino–antineutrino pair of that family. Experimentally all decays of the Z^0 are accounted for, and there is no room for such a decay.

The fact that there are three families and very likely no more precludes an interpretation in terms of bound states. This is different from the periodic system. We understand the occurrence of the elements as various bound states of electrons, protons and neutrons. It is a property of bound states that there is, normally, no end to such systems. You can always add protons (and neutrons). The fact that elements with more than 92 protons are unstable is not relevant in this context. It is for this reason that it is next to impossible to understand the three families as bound states of even more fundamental objects. Also, the (near) masslessness of the neutrinos is very hard to understand if they are to be bound states. How can we devise experimental methods to investigate such problems? No one knows. But as we discussed already above, there is this other curious object in the Standard Model, namely the Higgs. It has so far not been found, and also its theoretical properties are far from understood. In particular, its interactions with gravitation as predicted by the theory are very wrong and in contradiction with the Universe that we observe. Gravitation itself is not understood either. Perhaps all of these problems are related. Perhaps if we investigate the Higgs in great detail a clue to all the other problems may be found. This then is the hope for the future. Very likely the Higgs can be studied using machines under construction (the LHC at CERN, Geneva, a 14 TeV proton-proton collider) or

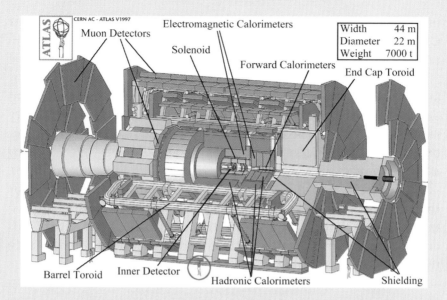

A drawing of the ATLAS detector being build for the LHC and scheduled to start operation in 2007. Note the size of the figure of a human being in the red circle.

The various detector elements are constructed to detect specific particles. The electromagnetic calorimeters are for the detection of photons and electrons. They consist of lead plates with liquid argon in between. Charged particles passing through the argon ionize the atoms, and the free electrons resulting from that are attracted (by means of an electric field) to electrodes and thus generate a current. The hadronic calorimeters are for pions, protons, etc. They consist of iron interleaved with plastic scintillator. Charged particles create light in the scintillator material which is recorded by light sensitive detectors. The inner detector contains thin slices of semiconducting material (the same as in computer chips) and a few hundred thousand hollow gas-filled 4 mm diameter pipes with a thin wire strung on their axis. For both detector types ionization is used to record the passage of a charged particle. The passage of muons is recorded in the outer shells of the ATLAS detector using the same principle.

The toroids and the solenoid are coils, generating a magnetic field. Each toroid is a system of 8 coils surrounding the beam pipe; one of the barrel toroid coils is indicated in the figure.

The group constructing the ATLAS detector consists of about 2000 scientists from more than 150 universities. The sociology of such a large collaboration is complicated, and that is certainly an unpleasant aspect of this large-scale experiment. One also needs people to fight for the money: the estimated cost is around US$320 million. The whole reminds one of NASA's efforts to send a probe to Mars.

machines yet on the drawing board (TESLA at DESY, Hamburg or CLIC at CERN, both electron-positron linear colliders). Perhaps, in another decade, Nature will unveil more of her wonderful secrets!

The reader may ask why in this book string theory and supersymmetry have not been discussed. String theory speculates that elementary particles are very small strings, and supersymmetry refers to the idea that corresponding to any particle there is another particle whose spin differs by $\frac{1}{2}$, at the same time invoking a large symmetry between the two types.[a]

The fact is that this book is about physics, and this implies that the theoretical ideas discussed must be supported by experimental facts. Neither supersymmetry nor string theory satisfy this criterion. They are figments of the theoretical mind. To quote Pauli: they are not even wrong. They have no place here.

---

[a]Particles of integral spin $(0, 1, 2, \ldots)$ are called bosons, and particles of half-integral spin $(\frac{1}{2}, \frac{3}{2}, \ldots)$ are called fermions. Supersymmetry is a boson-fermion symmetry.

Name Index[†]

[†]Page numbers followed by an asterisk refer to a page with a vignette.

Subject Index[†]

[†]Page numbers followed by an asterisk refer to a page with a vignette.

Photo Credits

Page 2, Planck: AIP[†] Emilio Segrè Visual Archives, W.F. Meggers Gallery of Nobel Laureates.

Page 9, Bohr: AIP Emilio Segrè Visual Archives.

Page 12: Rutherford: *Nature*, courtesy AIP Emilio Segrè Visual Archives.

Page 16, Maxwell: Original photograph in the possession of Sir Henry Roscoe, courtesy AIP Emilio Segrè Visual Archives.

Page 20, Dirac: Photo by A. Börtzells Tryckeri, courtesy AIP Emilio Segrè Visual Archives. E. Scott Barr and Weber Collections.

Page 28, Röntgen: Photo by Gen. Stab. Lit. Anst., courtesy AIP Emilio Segrè Visual Archives, Weber collection.

Page 28, X-ray hand: AIP Emilio Segrè Visual Archives, Lande collection.

Page 29, Marie and Pierre Curie: AIP Emilio Segrè Visual Archives, W.F. Meggers Gallery of Nobel Laureates

Page 36, Lorentz and Zeeman: © The Nobel Foundation.

Page 37, Thomson: © The Nobel Foundation.

Page 39, Millikan and Anderson: © The Nobel Foundation.

Page 40, Stückelberg: Courtesy Georges Stückelberg.

Page 48, Gell-Mann: Harvey of Pasadena, courtesy AIP Emilio Segrè Visual Archives.

Page 51, Pauli: CERN, courtesy AIP Emilio Segrè Visual Archives.

Page 56, Powell: © The Nobel Foundation.

[†]American Institute of Physics

Page 56, Perkins: Courtesy Donald Perkins.

Page 57. Occhialini and Blackett: Amaldi Archives, Dipartimento di Fisica, Università "La Sapienza", Rome, courtesy AIP Emilio Segrè Visual Archives.

Page 60, Uhlenbeck, Kramers and Goudsmit: AIP Emilio Segrè Visual Archives, Goudsmit collection.

Page 63, Richter and Ting: © The Nobel Foundation.

Page 65, Perl: © The Nobel Foundation.

Page 65, Zichichi: Courtesy Antonino Zichichi.

Page 72, Fermi: AIP Emilio Segrè Visual Archives, William G. Myers collection.

Page 74, Rubbia and van der Meer: © The Nobel Foundation.

Page 86, Heisenberg: AIP Emilio Segrè Visual Archives.

Page 87, Schrödinger: AIP Emilio Segrè Visual Archives.

Page 93, Born: © The Nobel Foundation.

Page 98, Cabibbo: Courtesy Nicola Cabibbo.

Page 105, Kobayashi: Courtesy Makota Kobayashi.

Page 105, Maskawa: Courtesy Toshihide Maskawa.

Page 110, Pais: Courtesy Nederlandse Natuurkundige Vereniging.

Page 127, Einstein: © The Nobel Foundation.

Page 141, C. Wilson: © The Nobel Foundation.

Page 142, Cherenkov, Frank and Tamm: © The Nobel Foundation.

Page 150, Glaser: © The Nobel Foundation,

Page 150, BEBC: Courtesy CERN.

Page 151, Bubble chamber picture: Courtesy CERN.

Page 156, Fukui: Courtesy CERN.

Page 156, Charpak: © The Nobel Foundation.

Page 158, Linde and Engelen: © M. Veltman.

Page 164, Hess: © The Nobel Foundation.

Page 164, Wulf: © Archiv der Norddeutschen Provinz SJ (ANPSJ) – Abt. 80 B 264.

Page 168, Cockroft and Walton: © The Nobel Foundation.

Page 170, Lawrence: Photograph by Watson Davis, Science Service, Berkeley National Laboratory, University of California at Berkeley, courtesy AIP Emilio Segrè Visual Archives, Fermi Film.

Page 170, cyclotron: Ernest Orlando Lawrence Berkeley National Laboratory, courtesy AIP Emilio Segrè Visual Archives.

Page 173, Touschek: Courtesy CERN Courier.

Page 177, aerial SLAC: Courtesy SLAC.

Page 177, drawing of TESLA: Courtesy DESY.

Page 182, Panofsky: AIP Emilio Segrè Visual Archives, *Physics Today* Collection.

Page 183, R. Wilson: AIP Emilio Segrè Visual Archives, *Physics Today* Collection.

Page 183, Adams: AIP Emilio Segrè Visual Archives.

Page 190, Schwartz: AIP Emilio Segrè Visual Archives, W.F. Meggers Gallery of Nobel Laureates.

Page 190, Pontecorvo: AIP Emilio Segrè Visual Archives, Segrè Collection.

Page 191, Lee and Yang: © The Nobel Foundation.

Page 195, Lederman and Steinberger: © The Nobel Foundation.

Page 197, Faissner, Krienen and Yang: Courtesy CERN.

Page 198, Shielding, bubble chamber and spark chamber: Courtesy CERN.

Page 202, Spark chamber event: Courtesy CERN.

Page 204, Bubble chamber event: Courtesy CERN.

Page 207, Faissner, von Dardel and Puppi: Courtesy CERN.

Page 212, Veltman and Bell: Courtesy CERN.

Page 215, Bernardini: Courtesy CERN.

Page 220, Alvarez: © The Nobel Foundation.

Page 221, Scanning table: Courtesy CERN.

Page 228, Nambu: Courtesy University of Chicago.

Page 245, Feynman: AIP Emilio Segrè Visual Archives, W.F. Meggers Gallery of Nobel Laureates and Weber collection.

Page 253, Lamb: © The Nobel Foundation.

Page 259, Crane: Courtesy Jens Zorn, University of Michigan.

Page 260, Garwin: Courtesy Richard Garwin.

Page 260, Telegdi: Courtesy Valentine Telegdi.

Page 274, Weinberg and Veltman: Courtesy The Royal Swedish Academy of Sciences.

Page 275, Glashow: Harvard University, courtesy AIP Emilio Segrè Visual Archives.

Page 275, 't Hooft: Courtesy The Royal Swedish Academy of Sciences.

Page 286, Higgs: Courtesy Peter Higgs.

Page 287, Brout: Courtesy Robert Brout.

Page 287, Englert: Courtesy François Englert.

Page 294, Bouchiat: Courtesy Claude Bouchiat.

Page 294, Iliopoulos: Courtesy John Iliopoulos.

Page 294, Meyer: Courtesy Philippe Meyer.

Page 299, Bjorken: Courtesy SLAC.

Page 305, aerial CERN and Weisskopf: Courtesy CERN.

Page 307, ATLAS: Courtesy CERN.